W9-DHK-516

Probing Plant Structure

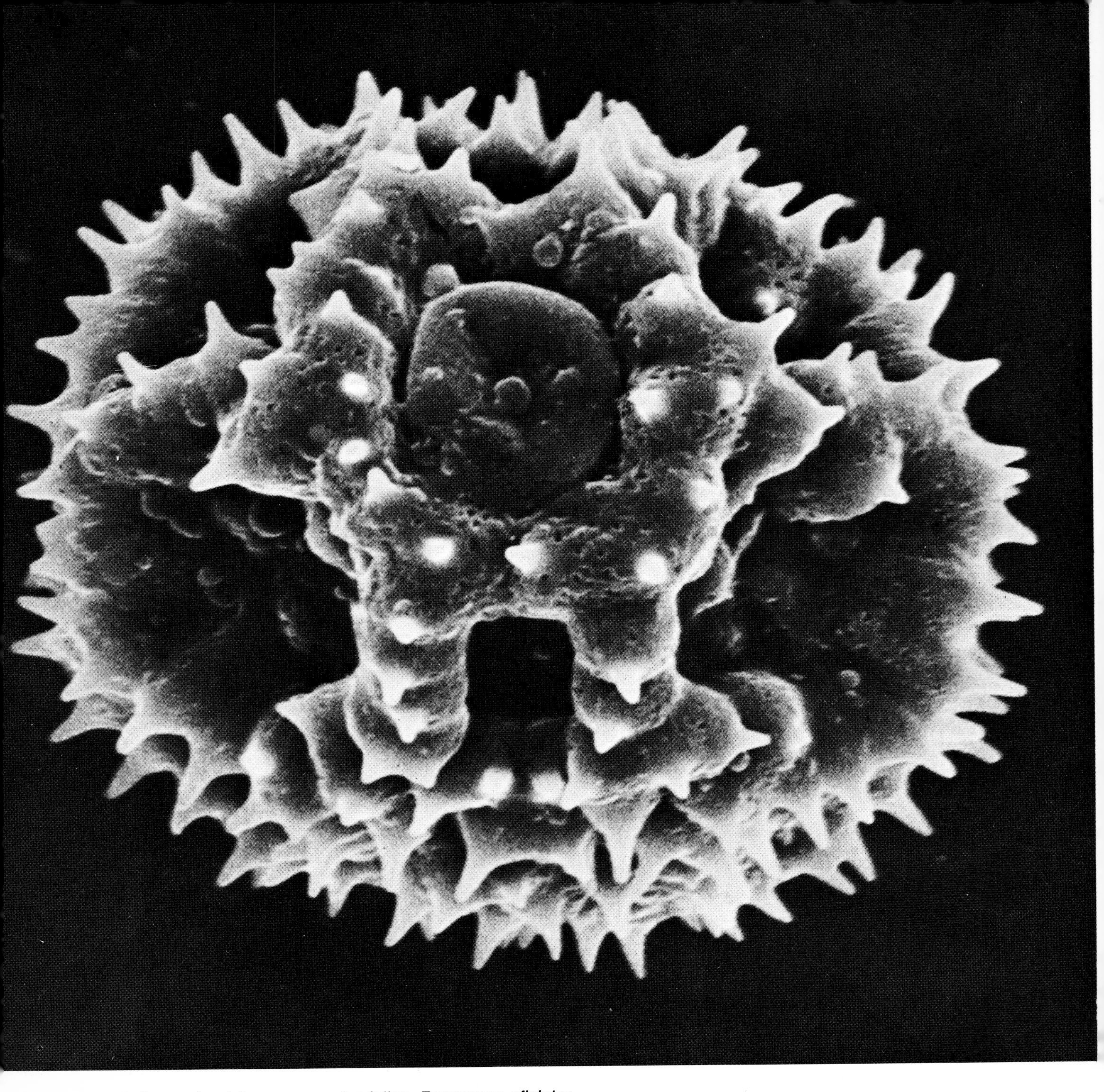

Pollen grain of the common dandelion, *Taraxacum oficiales.*

McGRAW-HILL BOOK COMPANY/NEW YORK, ST. LOUIS, SAN FRANCISCO

Probing Plant Structure

A Scanning Electron Microscope study of some anatomical features in plants and the relationship of these structures to physiological processes

John Troughton and Lesley A. Donaldson

Contents

Typeset by Typographical Services Ltd., Wellington

Printed by Dai Nippon Printing Co. (International) Ltd.
Hong Kong

Table of Plates

36. Broad Bean. Spongy mesophyll cell walls in a broad bean leaf.
37. Tomato. Palisade cells in a tomato leaf.
38. Broad Bean. Transverse view of a mature broad bean leaf.
39. Broad Bean. Transverse view of a young broad bean leaf.
40. Cotton. Transverse view of a leaf of cotton.
41. Cucumber. Transverse view of a leaf of cucumber.
42. Banana. Transverse view of a leaf of banana.
43. Pine. Transverse view of a pine needle.
44. Marram Grass. Transverse view of a leaf of marram grass.
45. Pineapple. Transverse view of a pineapple leaf.
46. *Atriplex hastata.* Transverse view of an *Atriplex hastata* leaf.
47. Saltbush. Transverse view of a saltbush leaf.
48. Saltbush. Transverse view of bundle sheath parenchyma cells in a leaf of saltbush.
49. *Gomphrena globosa.* Transverse view of a *Gomphrena globosa* leaf.
50. *Gomphrena globosa.* Longitudinal view of a *Gomphrena globosa* leaf.
51. *Gomphrena globosa.* Transverse bundle in a *Gomphrena globosa* leaf.
52. Portulaca. Chloroplasts in the bundle sheath cells of portulaca.

Chloroplasts

53. Portulaca. Chloroplasts aligned along the phloem in portulaca.
54. Portulaca. Chloroplasts in the bundle sheath cells of portulaca.
55. *Gomphrena globosa.* Chloroplasts in the bundle sheath cells of *Gomphrena globosa.*
56. Cucumber. Chloroplasts in a mesophyll cell of cucumber.

Xylem

57. Radish. The root tip of a radish seedling.
58. Maize. Transverse view of a prop root of maize.
59. Maize. Xylem vessels in a prop root of maize.
60. Cucumber. Xylem vessel in a cucumber root.
61. Cucumber. Xylem vessel in a cucumber stem.
62. Cotton. Xylem vessels in a cotton petiole.
63. Cucumber. Xylem vessels in a cucumber stem.
64. Cucumber. Spiral secondary thickenings in a xylem vessel from a cucumber stem.
65. Cucumber. Scalariform secondary thickenings in a xylem vessel from a cucumber stem.
66. Cucumber. Reticulate secondary thickenings in a xylem vessel from a cucumber stem.
67. Cucumber. Reticulate secondary thickenings in a xylem vessel from a cucumber stem.
68. Maize. Reticulate secondary thickenings in a xylem vessel from a maize leaf.

The Apical Meristem

69. Wheat. The apical meristem of Arawa wheat at the late vegetative stage.
70. Wheat. The apical meristem of Arawa wheat at the late vegetative stage.
71. Wheat. The apical meristem of Raven wheat in the reproductive stage.
72. Wheat. The apical meristem of Raven wheat in the reproductive stage.
73. Wheat. The apical meristem of Raven wheat in the reproductive stage.
74. Broad bean. The floral apex of broad bean.

The Flower

75. Alyssum. The flower of alyssum.
76. Alyssum. Cells on the surface of a petal of alyssum.
77. Paspalum. The flower of paspalum.
78. Saltbush. Anthers from a saltbush flower.
79. Cotton. Anthers, covered in pollen grains, from a cotton flower.
80. Maize. Transverse view of an anther from a maize flower.
81. Maize. Transverse view of a locule from a maize anther.
82. Wheat. Anther with pollen grains in wheat.
83. Wheat. Surface features of a wheat anther.
84. Cotton. Pollen grain of cotton.
85. Maize. Pollen grain of maize.
86. Paspalum. Pollen grain of paspalum.
87. Pine. Pollen grain of pine.

Fertilisation and Seed Development

88. Wheat. The flower of wheat with palea removed.
89. Alyssum. Gynoecium from a flower of alyssum.
90. Daphne. Stigma from a flower of daphne.
91. Primrose. Stigma from a flower of primrose.
92. Cotton. Pollen grains on the stigma of cotton flower.
93. *Gomphrena globosa.* Stigma from a flower.
94. *Gomphrena globosa.* Pollen grains on the stigma of saltbush.
95. Paspalum. Pollen grain on the stigma of paspalum.
96. Wheat. Pollen grain on the stigma of wheat.
97. *Felicia ameloides.* Fruit and receptacle of *Felicia ameloides.*
98. *Felicia ameloides.* Single receptacle of *Felicia ameloides.*

Seeds

99. Burr clover. Seed case of burr clover.
100. Saltbush. Seed case of saltbush.
101. *Galinsoga parviflora.* Seed of *Galinsoga parviflora.*
102. *Galinsoga parviflora.* Pappus on a *Galinsoga parviflora* seed.
103. *Epilobium sp.* Seed of an *Epilobium sp.*
104. *Epilobium sp.* Seed of an *Epilobium sp.*
105. *Silene cucubalus.* Seed of bladder campion.

List of Species

Common Name	Botanical Name
aloe	*Aloe variegata*
alyssum	*Lobularia maritima*
	Atriplex hastata
banana	*Musa sp.*
bladder campion	*Silene cucubalus*
broad bean	*Vicia faba*
burr clover	*Medicago hispida*
carnation	*Dianthus sp.*
cotton	*Gossypium hirsutum*
cucumber	*Cucumis sativus*
daphne	*Daphne sp.*
epilobium	*Epilobium sp.*
	Fellicia ameloides
	Galinsoga parviflora
gomphrena	*Gomphrena globosa*
maize	*Zea mays*
marram grass	*Ammophila arenaria*
New Zealand Christmas Tree	*Metrosideros excelsa*
Norfolk Island hibiscus	*Lagunaria patersonii*
oleander	*Nerium oleander*
opuntia	*Opuntia sp.*
paspalum	*Paspalum sp.*
pine	*Pinus*
pineapple	*Ananas comosus*
portulaca	*Portulaca sp.*
radish	*Raphanus sativus*
saltbush	*Atriplex spongiosa*
tomato	*Lycopersicon esculentum*
Townsville stylo	*Stylothanthes humilis*
wheat	*Triticum aestivum*

Emphasis is given throughout this book to aspects of photosynthesis involving carbon metabolism, and it is possible to group the plants given in this list into three categories based on the mode of carbon fixation and transformation.

Group I: C_3 type plants. Ribulose -1:5-diphosphate carboxylase (Ru DP carboxylase) is the enzyme involved in the initial event of CO_2 fixation in the light. A three carbon compound is produced. Examples of C_3 type plants are wheat, tomato, cucumber, cotton and broad bean.

Group II: C_4 type plants. C_4 type plants have phosphoenolpyruvate carboxylase (PEP carboxylase) as the enzyme involved in the primary carboxylation event. This fixation of CO_2 occurs in the light, and four carbon compounds, malate and aspartate, are produced. These acids are then immediately metabolised in the light to produce more complex carbon compounds by reactions similar to those in C_3 type plants. Examples of C_4 type plants used in this book include maize, paspalum, saltbush, portulaca and *Gomphrena globosa.*

Group III: Crassalacean Acid Metabolism (CAM). This type of carbon metabolism is similar to that of C_4 type plants in that PEP carboxylase is the enzyme involved in the initial carboxylation event. However, this carboxylation reaction takes place in the dark and the malate that is produced is not converted to other carbon compounds until the subsequent light period. In C_4 type plants the two sets of reactions, the C_4 and C_3 type, are spatially separated by being localised in different cells while in CAM plants these two sets of reactions involve temporal separation. CAM type plants are normally succulents but some succulents are C_3 type plants. Opuntia, pineapple and *Aloe variegata* are three species exhibiting crassalacean acid metabolism.

Preface

The complexity of a multicellular organism such as a plant often directs research into narrow and widely separated subjects, and teaching tends to follow and reinforce these arbitrary divisions. Yet these divisions hinder an understanding and appreciation of plants as highly integrated organisms in which one structure is related to another, and structure is related to function. One of the objectives of this book has been to bring together a limited number of aspects of plant anatomy and physiology that are important to plant growth. The problem of describing anatomical features of plants has been overcome by the use of photographs taken on the Scanning Electron Microscope (SEM). These photographs require a minimum of interpretation so that the descriptive part of the text has been reserved to explain the physiological significance of the structures that are illustrated.

The authors have been forced to limit both the number of components in the plant that were investigated and also the number of illustrations of any one feature that could be included in the book. In its present form the book concerns some aspects of three basic processes in plants: carbon dioxide uptake, water transport and reproduction. It is intended that this book should be used to complement existing textbooks but at the same time it will be useful for practical classes in biology where the photographs will assist in the interpretation and understanding of fresh material or slides.

Probing Plant Structure is primarily directed to students in biology who will be attempting to understand how a plant functions. However, this does not exclude the usefulness of the photographs to researchers, teachers and advanced students who, it is hoped, will obtain a fresh appreciation of plant anatomy. To many people the photographs will reinforce an understanding of anatomy they already hold, but it is hoped that all readers are stimulated into further investigations into the ecology, anatomy, physiology, physics and chemistry of plants.

Acknowledgments

The authors record their sincere appreciation for assistance from many people and especially for the encouragement and guidance given to the project by Dr M. C. Probine, Director of the Physics and Engineering Laboratory, Department of Scientific and Industrial Research.

The project was conducted with the support of the Electron Microscope section at the Physics and Engineering Laboratory. In particular, Mr W. S. Bertaud and Dr N. E. Flower freely gave invaluable assistance in maintaining the SEM in excellent working condition. Mr L. M. Adamson was responsible for the preparation of the prints for publication and Mr G. D. Walker and Mrs S. M. O'Kane gave assistance in some aspects of the work.

Mr R. Braizier, New Zealand Geological Survey, DSIR, and Mrs K. A. Card, Physics and Engineering Laboratory, DSIR, also assisted with aspects of the preparation of photographs. Mrs Card was also responsible for the growing of the plant material. There were several sources of seed and plant material and the co-operation of the following people or organisations in supplying samples is appreciated: the New Zealand Department of Agriculture Seed Testing Station, Palmerston North; Dr J. E. Begg, Division of Land Research, CSIRO, Canberra; Mr V. Southwell, Research School of Biological Sciences, Australian National University, Canberra; Parks and Reserves Department, Lower Hutt City Corporation, Lower Hutt; and Mr G. K. Cotterill, Physics and Engineering Laboratory, DSIR, Lower Hutt.

Several people were consulted on aspects of the contents of this book and the authors appreciate the advice given by Mr B. A. Meylan, Physics and Engineering Laboratory, DSIR, Lower Hutt; Mrs G. J. Smart, Hutt Valley High School, Lower Hutt; Dr W. B. Silvester and Professor L. H. Millener, Botany Department, University of Auckland, Auckland; and Professor T. M. Morrison and Professor R. H. M. Langer, Lincoln College, Christchurch, and Professor H. D. Gordon, Botany Department, Victoria University of Wellington.

Introduction

The Scanning Electron Microscope is an instrument that can be used to observe the surface of objects and in several respects it complements and extends the capabilities of the more commonly used optical and transmission electron microscope. The main advantage of the SEM technique is that it achieves a depth of focus about 300 times better than the optical microscope and this gives a three dimensional appreciation of the object. Another advantage of the instrument is that the resolution (20 nanometers, nm) is superior to the light microscope (500nm) although it is inferior to the transmission electron microscope (about 1nm).

The SEM operates by focussing a beam of primary electrons into a fine spot (about 10 nm in diameter) which scan the specimen. When the beam strikes the specimen, secondary electrons are emitted. These electrons are collected, allowed to impinge on a scintillator and the photons released are detected by the photo-cathode of the photomultiplier. The signal from the photomultiplier is then used to modulate the brightness of the cathode ray tube. The image so produced can be viewed and has a three dimensional appearance because of the contrast that is produced in different parts of the specimen, depending on the number of secondary electrons that are emitted. A permanent record can be obtained by photographing the image.

Most of the specimens used during the investigations were obtained from plants grown in controlled environment cabinets at the Physics and Engineering Laboratory of the DSIR, Lower Hutt. Some samples were obtained from plants grown in a glasshouse under the supervision of the Parks and Reserves Department, Lower Hutt City Corporation, while a few samples came from plants growing naturally outdoors.

Specimens used in scanning electron microscopy can be prepared more quickly than specimens for transmission electron microscopy, because embedding and staining procedures are not required. However it is necessary to prevent charge build-up on the specimens and this is avoided by coating the specimen and the stub on which the specimen is mounted, with a thin layer (10nm) of gold-palladium. This metal coating provides a low-resistance leakage path between the specimen and ground. This preparation technique has been used at the Physics and Engineering Laboratory for a wide variety of materials which have included wool fibres, pollen grains, leather, clays, metal surfaces, seeds, insects, wood structure, pulp, paper and ion exchange resins. It is not always necessary to coat specimens with metal and some of the photographs in this book are of uncoated material, but this restricts the resolution that can be obtained.

Biological specimens are generally more difficult to prepare because they are in a highly hydrated state. Most of the materials used to obtain the photographs in this book were prepared by freeze-drying. The samples were rapidly frozen in freon at —130°C and then transferred to liquid nitrogen. Subsequently the specimens were freeze-dried at —70°C for about 20 hours and then coated with a thin layer of carbon and gold-palladium. To observe the internal structure of plant tissue it was necessary to freeze-fracture the samples while they were in liquid nitrogen and prior to freeze-drying.

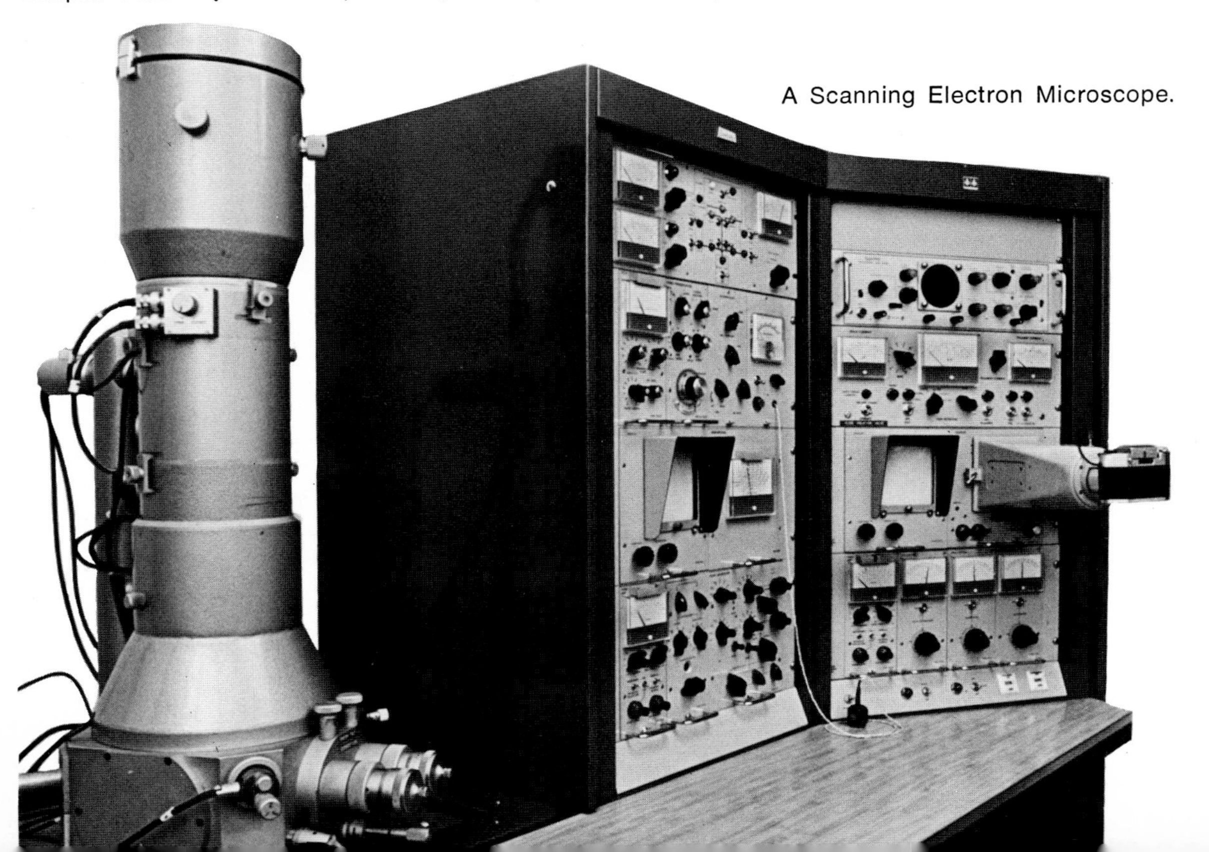

A Scanning Electron Microscope.

Plate 1 x1670	Leaf Surface

Hairs on the lower surface of a leaf of the New Zealand Christmas Tree *(Metrosideros excelsa;* Maori name. pohutukawa). The leaf specialises in photosynthesis and epidermal structures on the surface interfere with photosynthesis because they restrict the rate of supply of CO_2 to the cells in the leaf. This tree is indigenous to New Zealand and grows in sandy soils and on cliff faces in coastal areas. Both these ecological sites are subjected to wide variations in the soil water content and the profuse development of hairs on the leaf is a method of conserving plant water by restricting the rate of water loss from the leaf.

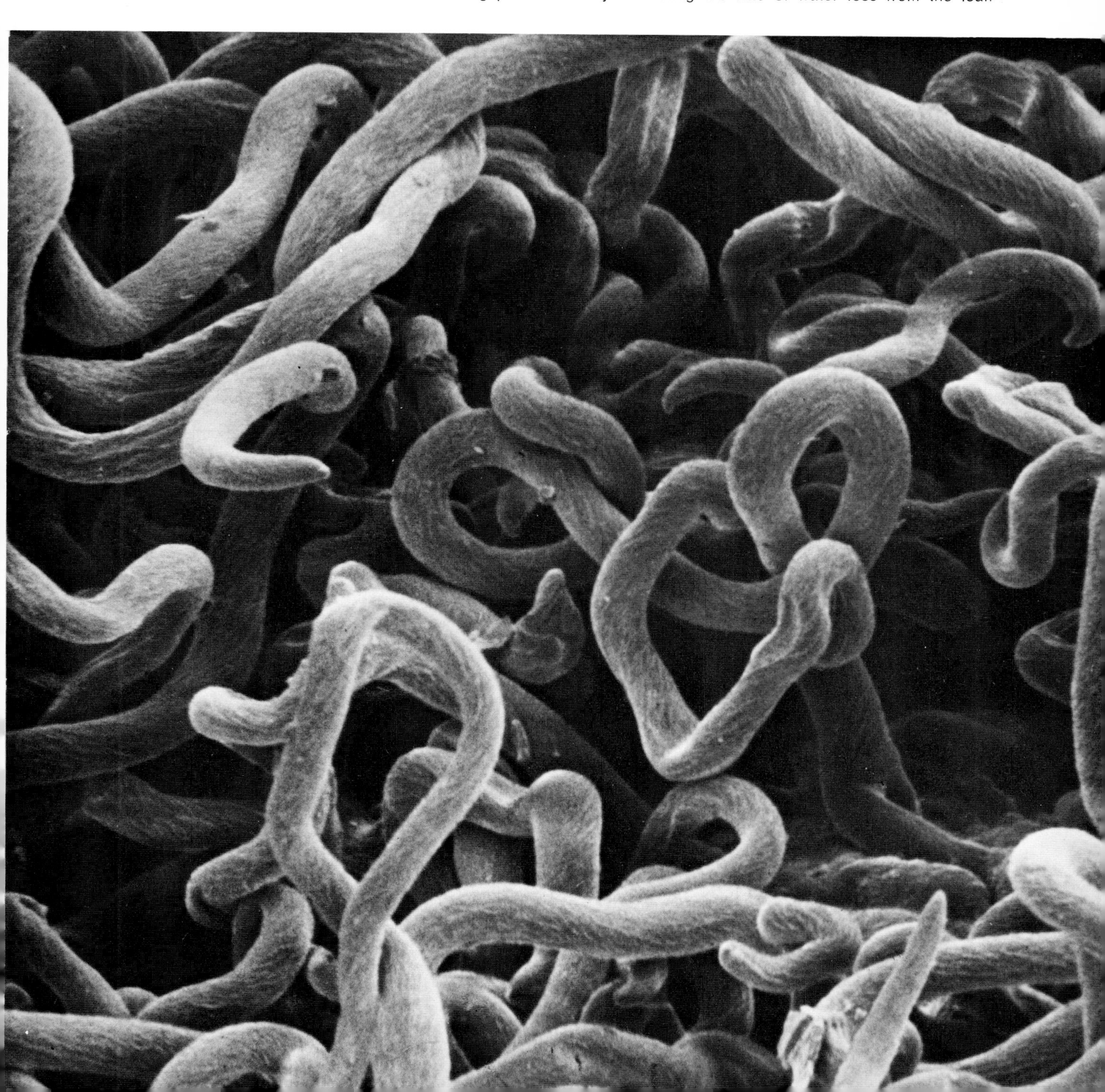

Plate 2 x420 Leaf Surface

Stellate hairs on a leaf of Norfolk Island hibiscus *(Lagunaria patersonii).* This leaf appears smooth to the naked eye but as this photograph shows the epidermis is completely covered by hairs. The hair on a hibiscus leaf is attached to the plant by a foot embedded in the epidermis. These hairs consist of several cells, fused together towards the base and separated at the margin whereas the hairs on the New Zealand Christmas Tree are in contrast to these stellate hairs because they are single celled.

Plate 3 x252 — Leaf Surface

Hairs on the lower surface of a cucumber *(Cucumis sativus)* **leaf.** There are fewer hairs on the leaf surface of cucumber compared with the previous examples and although they are large they are unlikely to significantly affect CO_2 uptake or water vapour loss from the leaf. They are multicellular hairs and on many plants such hairs are living. Stomata are present on the leaf surface and the relative sizes of stomata and hairs may be noted. The upper surface of a cucumber leaf has fewer, although larger, hairs than those shown in this photograph.

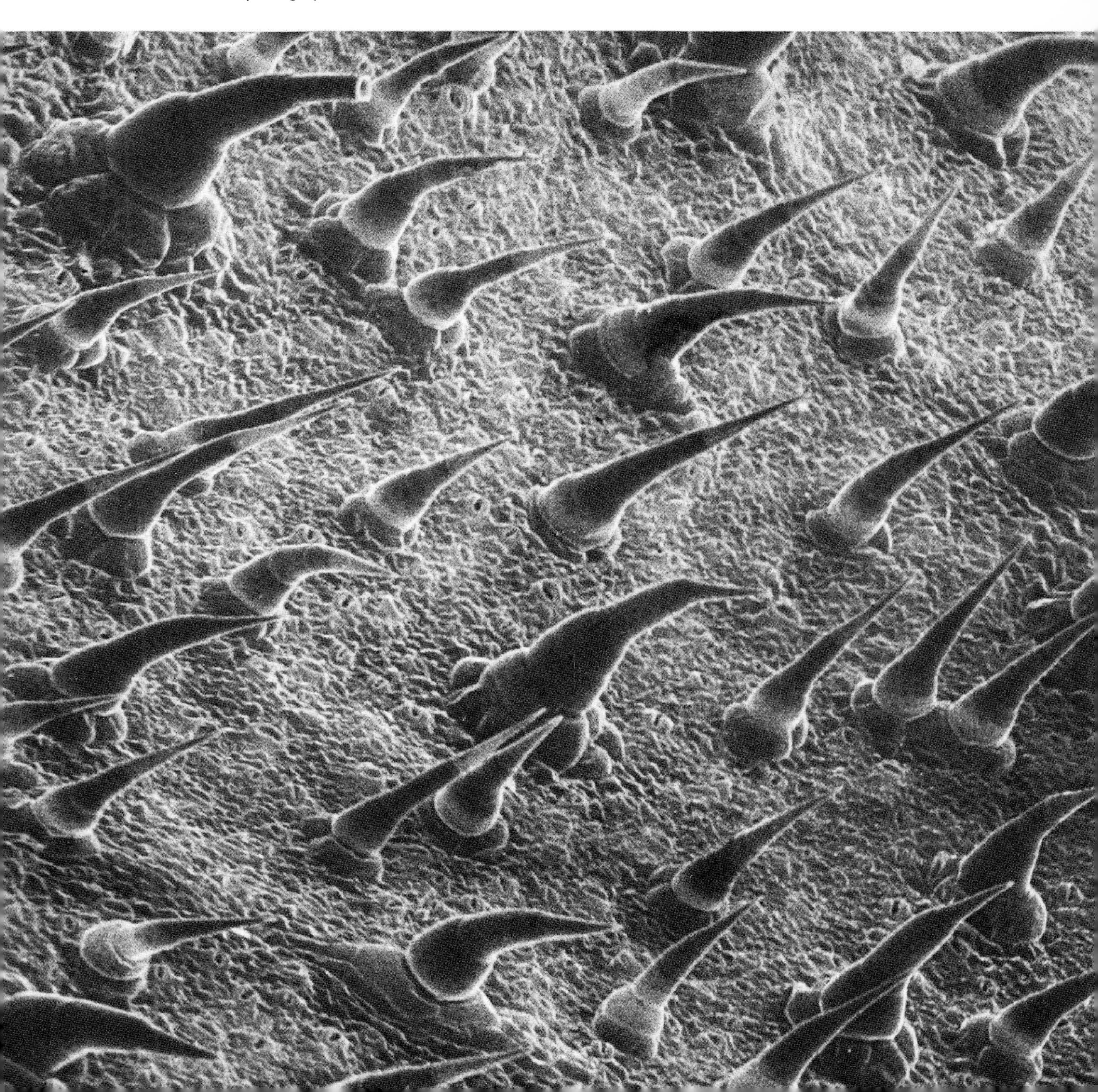

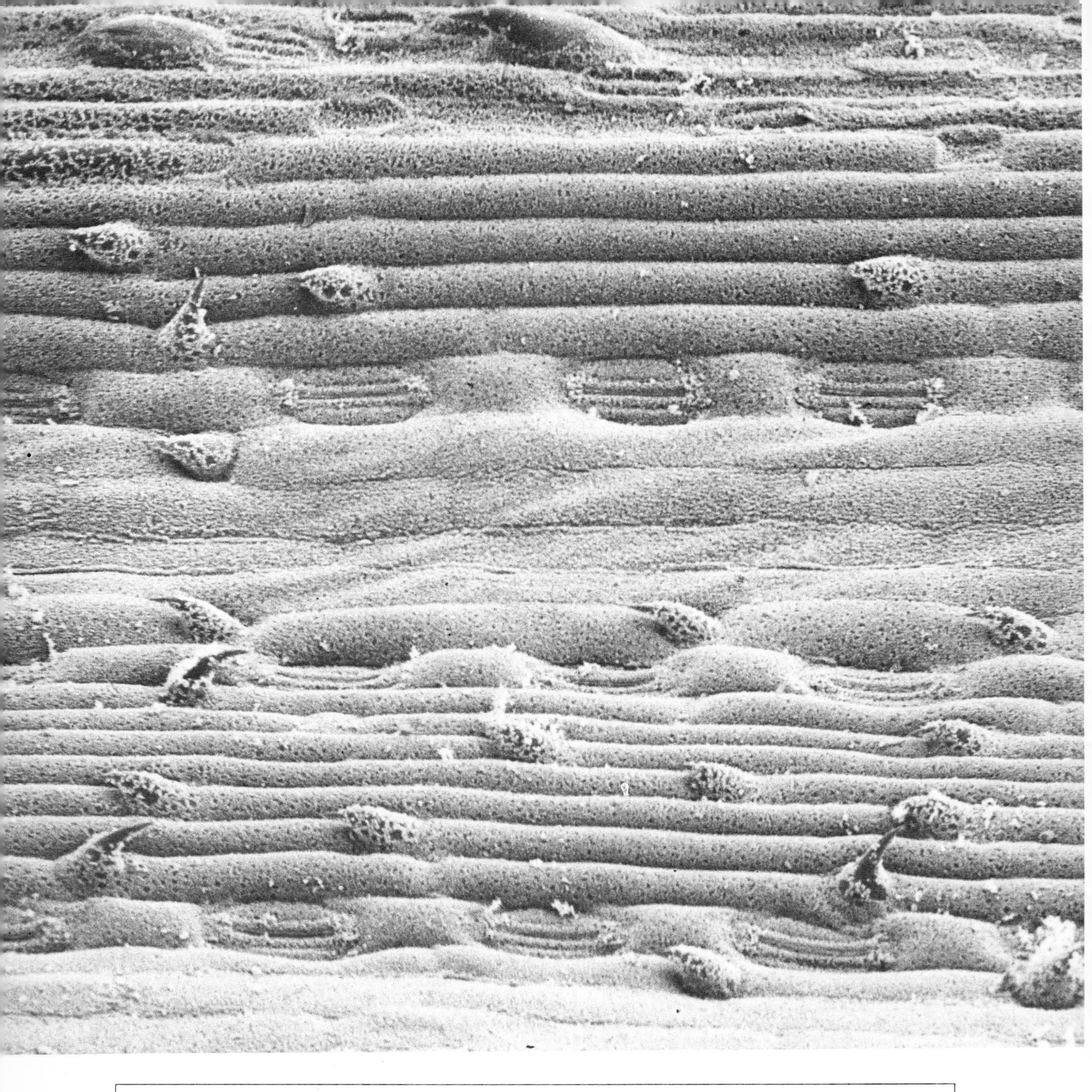

Plate 4 x540	Leaf Surface

Upper surface of the flag leaf of wheat *(Triticum aestivum;* variety Arawa). The only epidermal structures present on Arawa wheat leaves are small spines which are unlikely to have any physiological significance. The flag leaf is the leaf nearest to the wheat ear and is important because about 40% of the carbon that appears in the wheat grain comes from photosynthesis in the flag leaf. The photograph also shows that wax covers the surface of the leaf and it would be difficult for chemical sprays to enter the leaf unless the effect of the wax was overcome. The cuticle in wheat would be impermeable to water because it is impregnated with wax. The unwettable nature of a surface such as this one would also assist the leaf in resisting invasion from pathogens. Wax also covers the surface of the stomata, which can be seen in parallel rows along the length of the leaf.

Plate 5 x190 Leaf Surface

Surface of the stem of a cactus plant. Hairs or spines cover the leaf and stem surfaces of many cactus plants. These hairs are clustered together and meet over the top of the stomata (S). In this position, these hairs may assist in reducing transpiration. A feature of xerophyte plants is their ability to resist desiccation because of a cuticle impermeable to water and relatively few stomata on the plant surface.

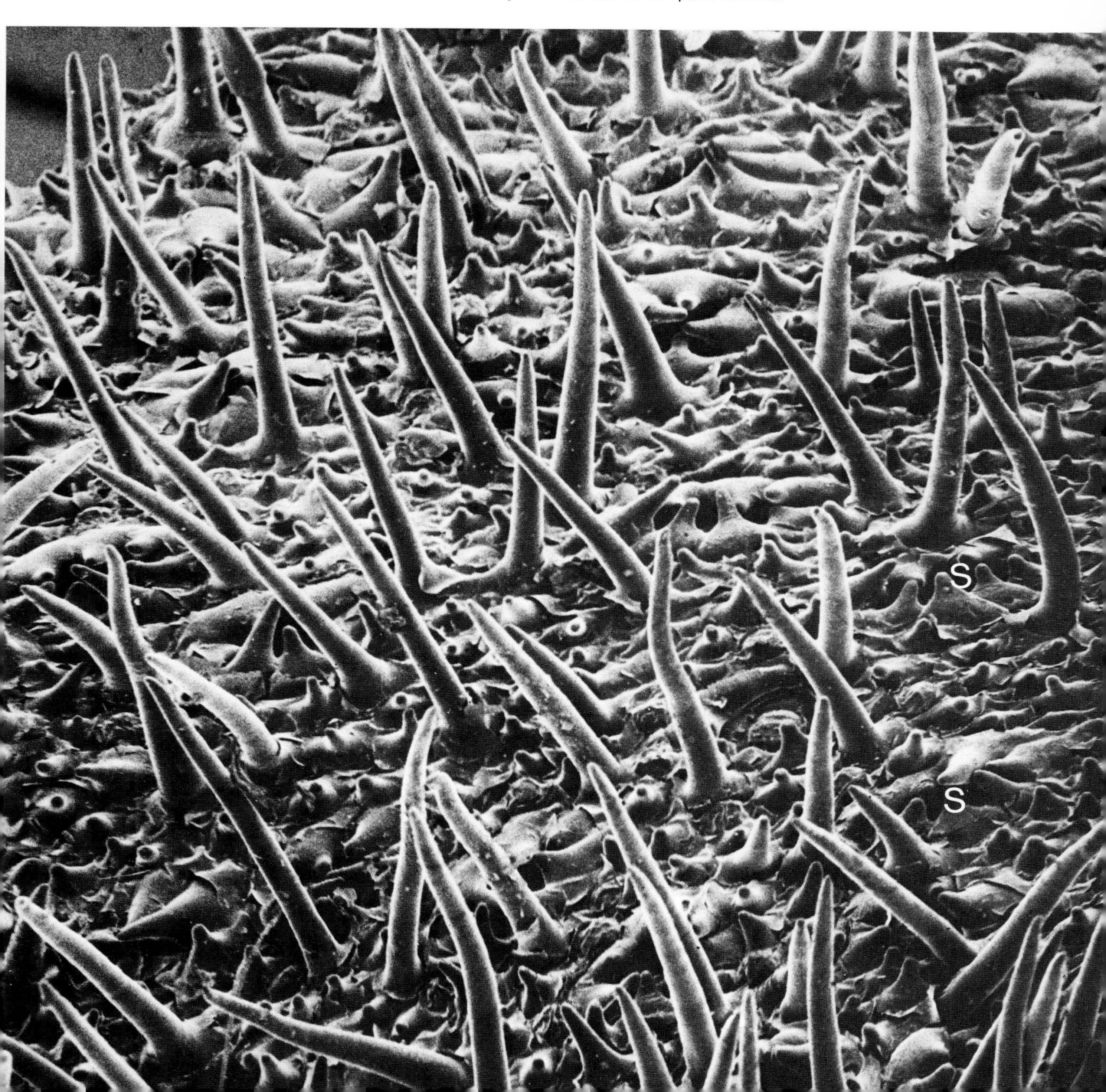

Plate 6 x620 — Leaf Surface

Spines on the stem of an *Opuntia* species. The spines on this cactus are not visible to the naked eye, but their presence is quickly established when touched. It has been suggested that spines such as these prevent insects from damaging the cuticle thereby causing the cactus plant to lose water.

Plate 7 x460 Leaf Surface

Salt bladders on the leaf of saltbush *(Atriplex spongiosa)*. Saltbush is indigenous to Australia and grows in arid areas, which often also have saline soils. This plant is a salt tolerant species because it has developed a mechanism to control the N+ and Cl— ion concentration of its tissues. The epidermal bladders on the surface of the aerial parts of the plant are specialised cells that accumulate salt. Salt from the leaf tissues is transferred through the small stalk cell (S) and into the balloon-like bladder cell (B). As the leaf ages the salt concentration in the cell increases and eventually the cell bursts or falls off the leaf, releasing the salt outside the leaf. The bladder cell can control the salt concentration and osmotic pressure in the leaf at normal levels while in the cell the osmotic pressure may be negative by several hundred atmospheres.

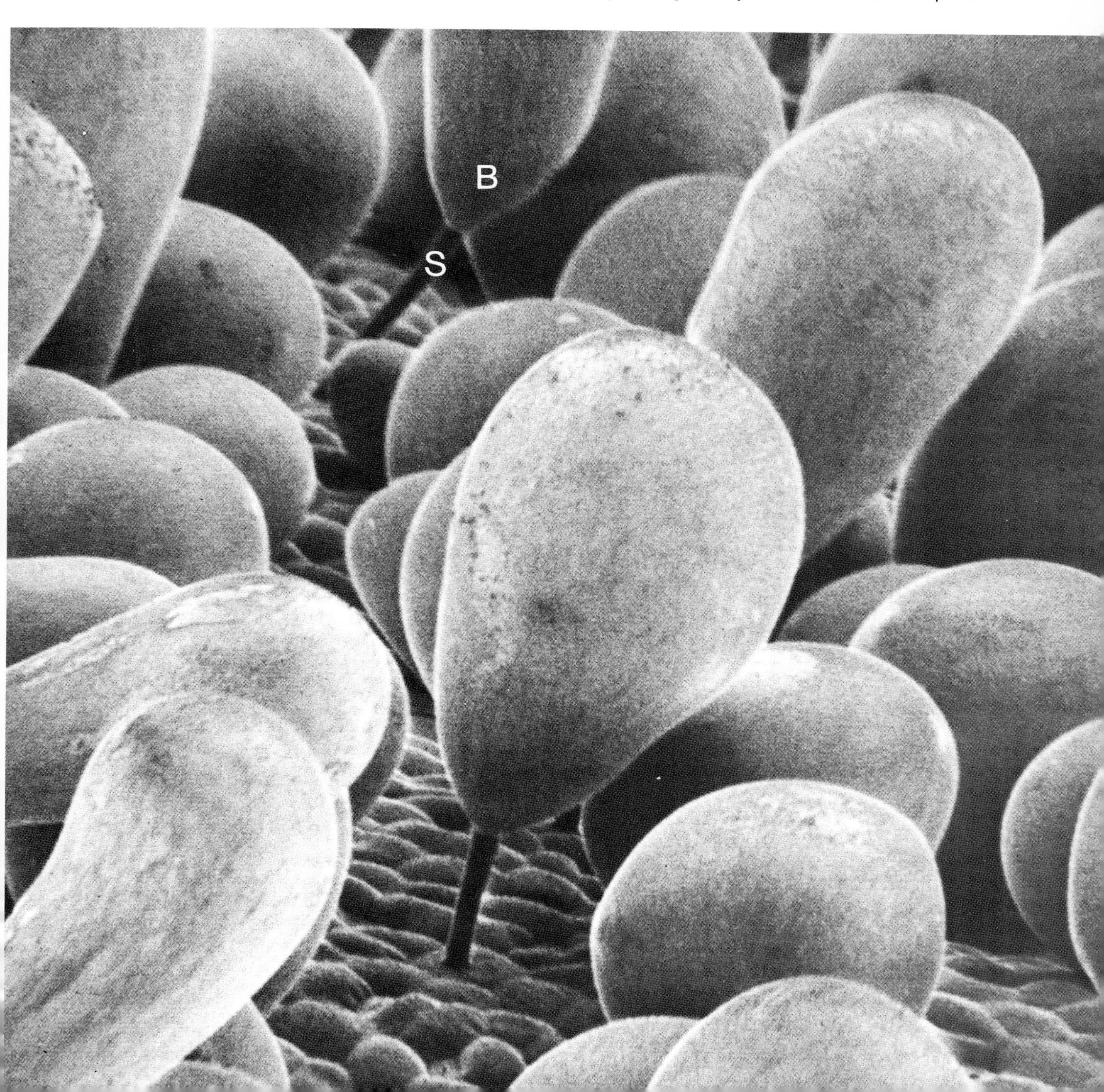

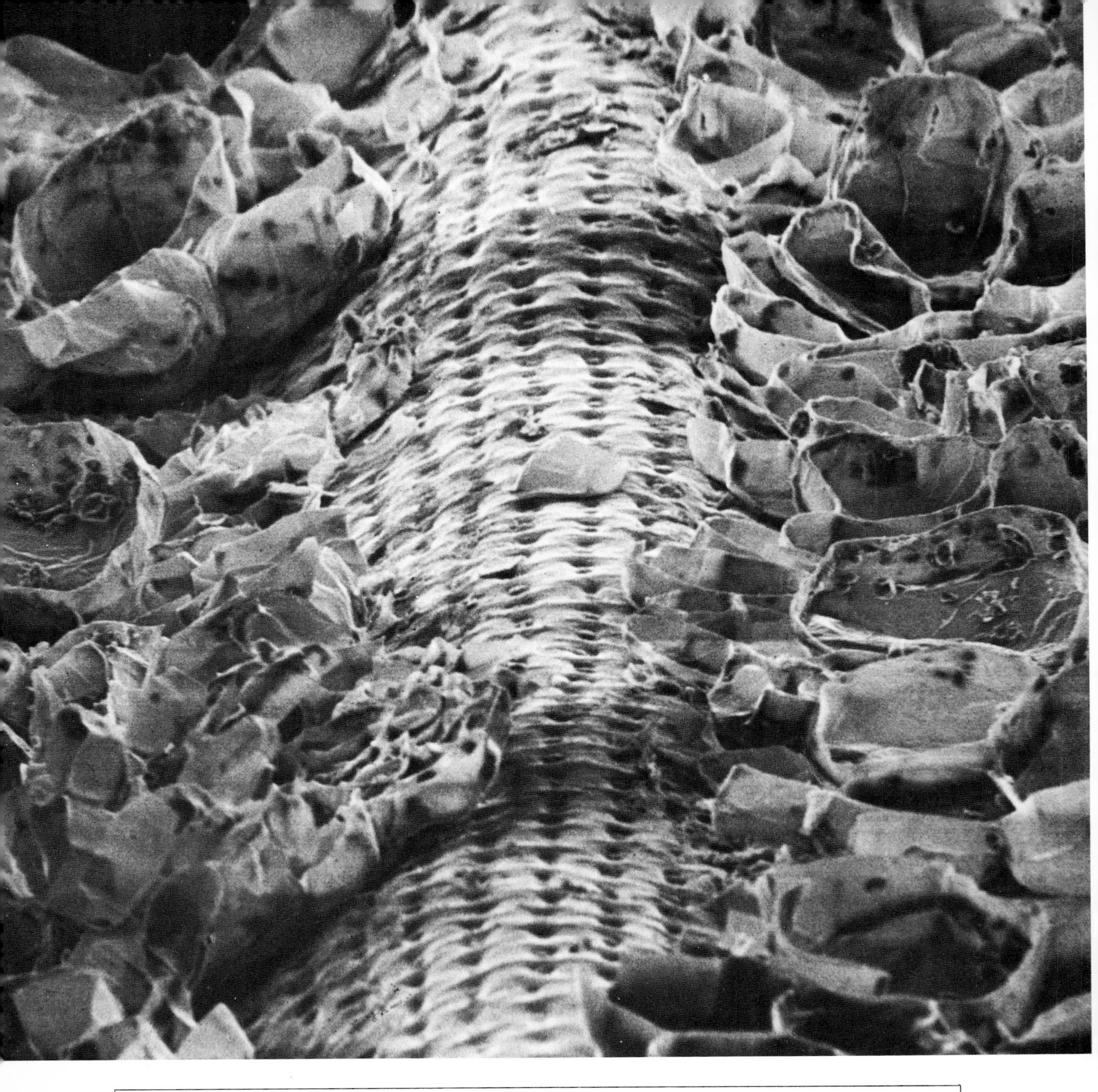

Plate 8 x570 Leaf Surface

Scales on the lower surface of a leaf of pineapple *(Ananas comosus).* The leaf surface is ridged and in this example the scales are localised in the troughs. The stomata are located beneath the scales and are therefore not visible. The indentations in the cells on the ridge indicate that crystal structures are present in the cells. When the scales are wet they allow water to enter the leaf, but when partially dehydrated the scales close together to prevent transpiration from the leaf.

Plate 9 x3900 **Leaf Surface**

Wax on the leaf surface of carnation. Carnation *(Dianthus sp.)* is another common plant that has a prolific layer of wax on the cuticle, as shown in this photograph. The wax is synthesised within the epidermal cells and is forced through channels in the cell wall and cuticle and deposited on the leaf surface. The structure of wax on the surface can be as thin flakes, plates, rodlets or rods. When the wax is in the form of rodlets or rods it is visible to the naked eye as a bluish 'bloom', which can be easily rubbed off the leaf.

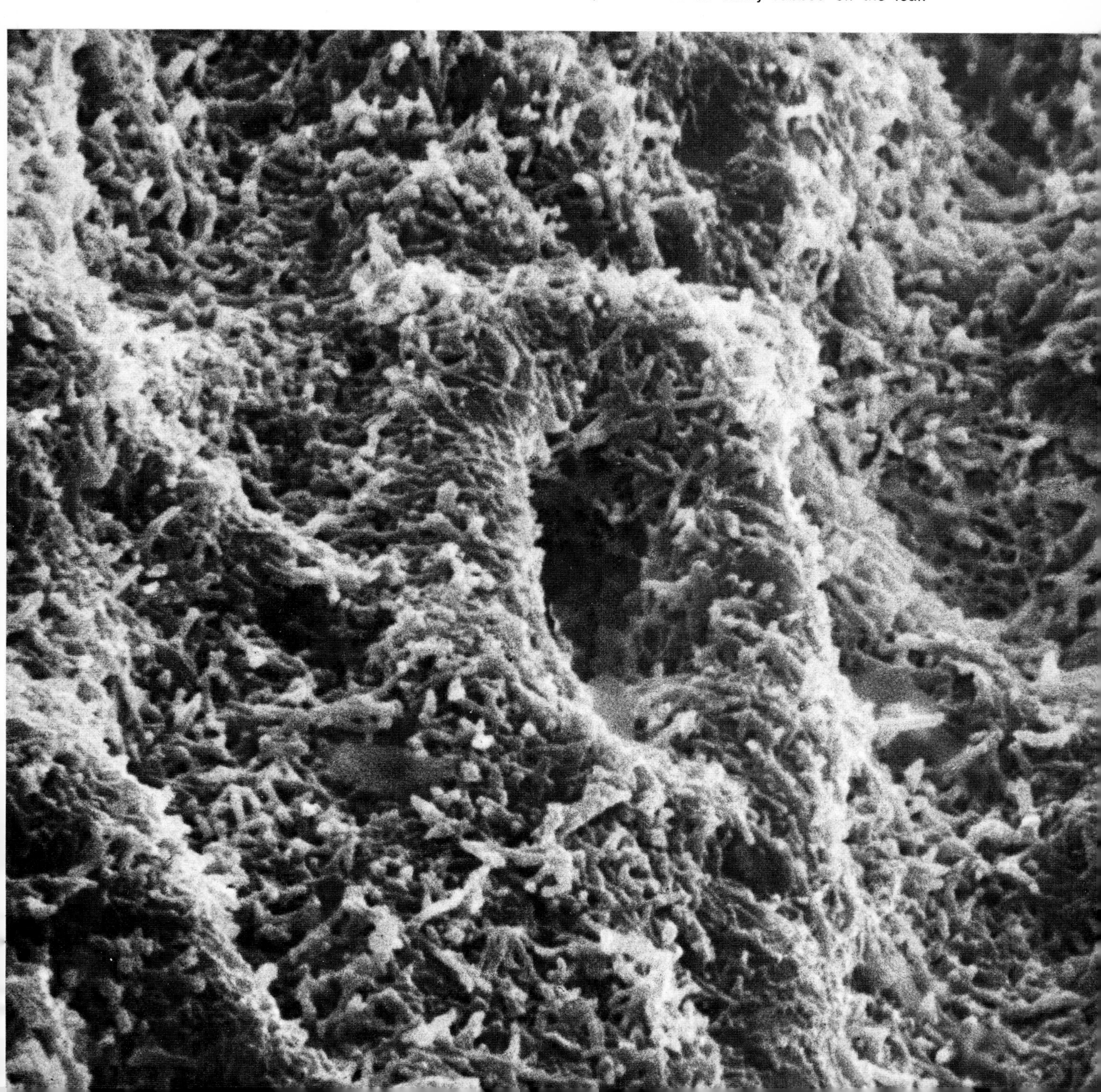

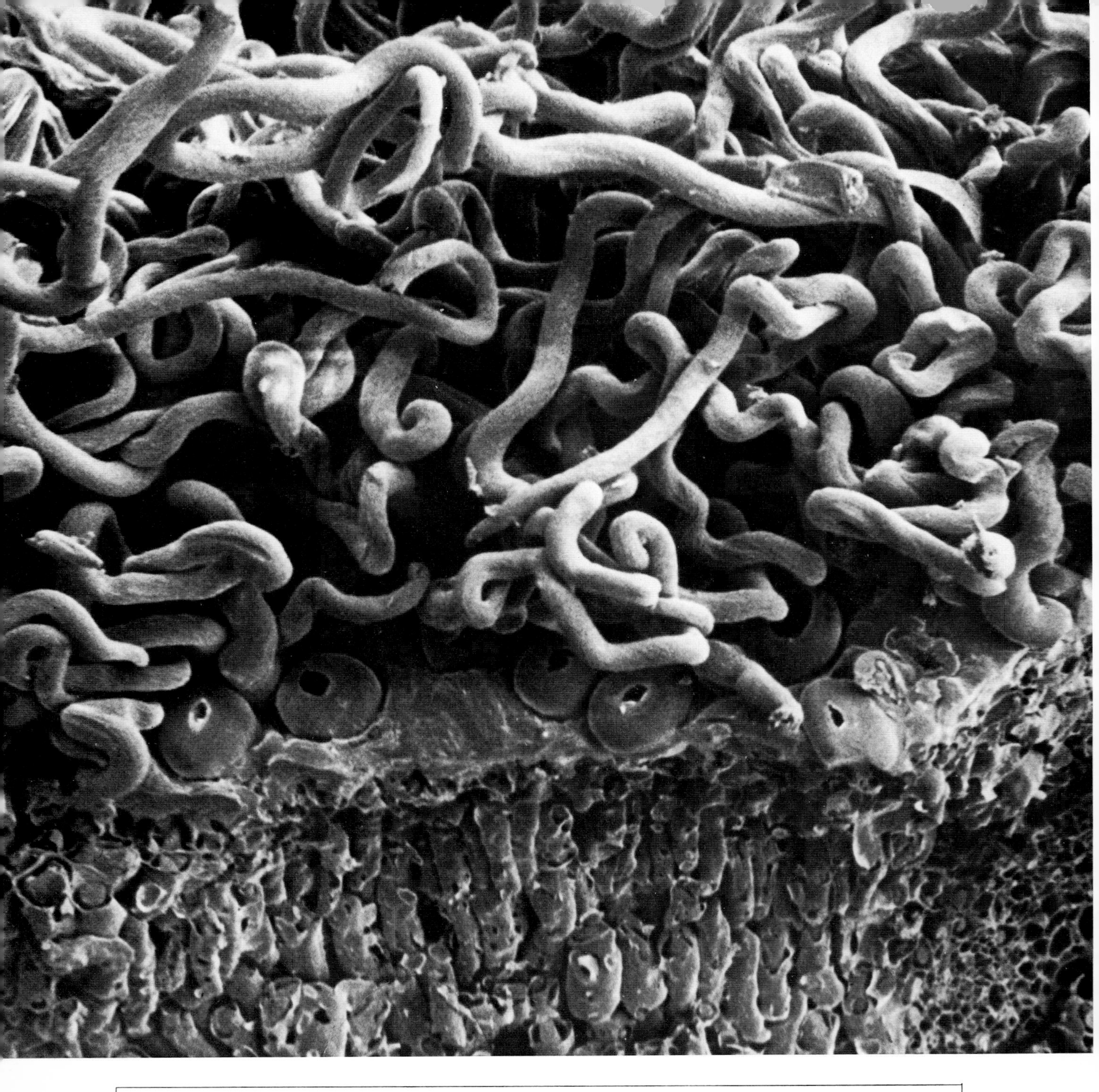

Plate 10 x700 — Leaf Surface

Transverse view of a leaf of the New Zealand Christmas Tree. Numerous stomata can be seen as part of the epidermis, which is at the base of the thick layer of hairs. In this position the stomata are remote from the air outside the leaf and the hairs interfere with the carbon dioxide moving into the leaf during photosynthesis. The upper surface of the same leaf is almost free of hairs but CO_2 cannot enter the leaf through that surface because there are few stomata present and there is a very thick cuticle.

Plate 11 x900	Stomata

Stomata on the upper surface of a leaf of maize *(Zea mays)*. The stomata on a maize leaf are also arranged in parallel rows but along the length of the leaf. The epidermal cells can also be seen, with the margin between two cells appearing as a wavy line. These cells are long and narrow. In warm regions maize has a high growth rate which may be partially due to the many large stomata on the leaf. However, there are several other contributing factors to the growth rate and of these the photosynthetic mechanisms are especially important.

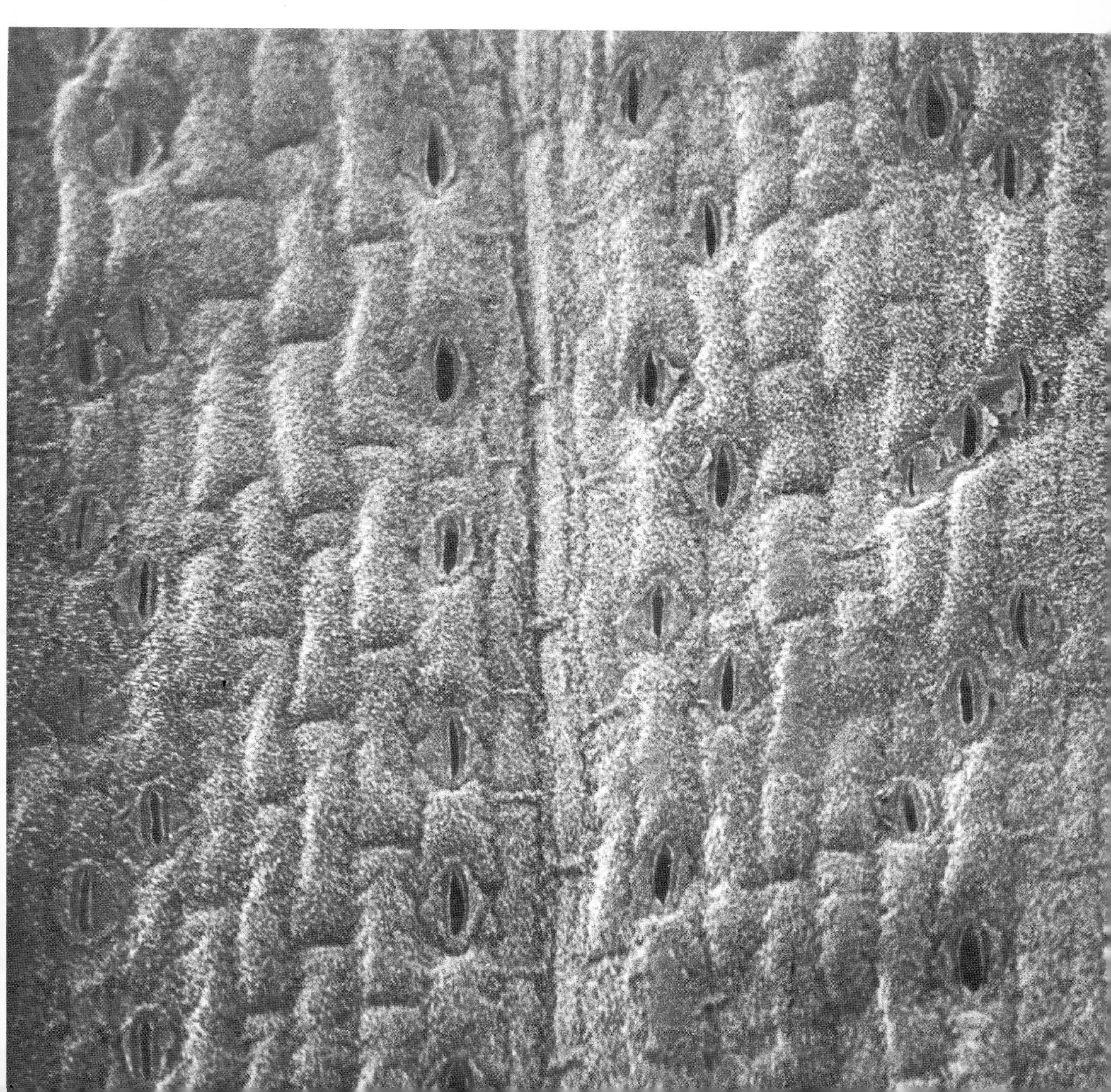

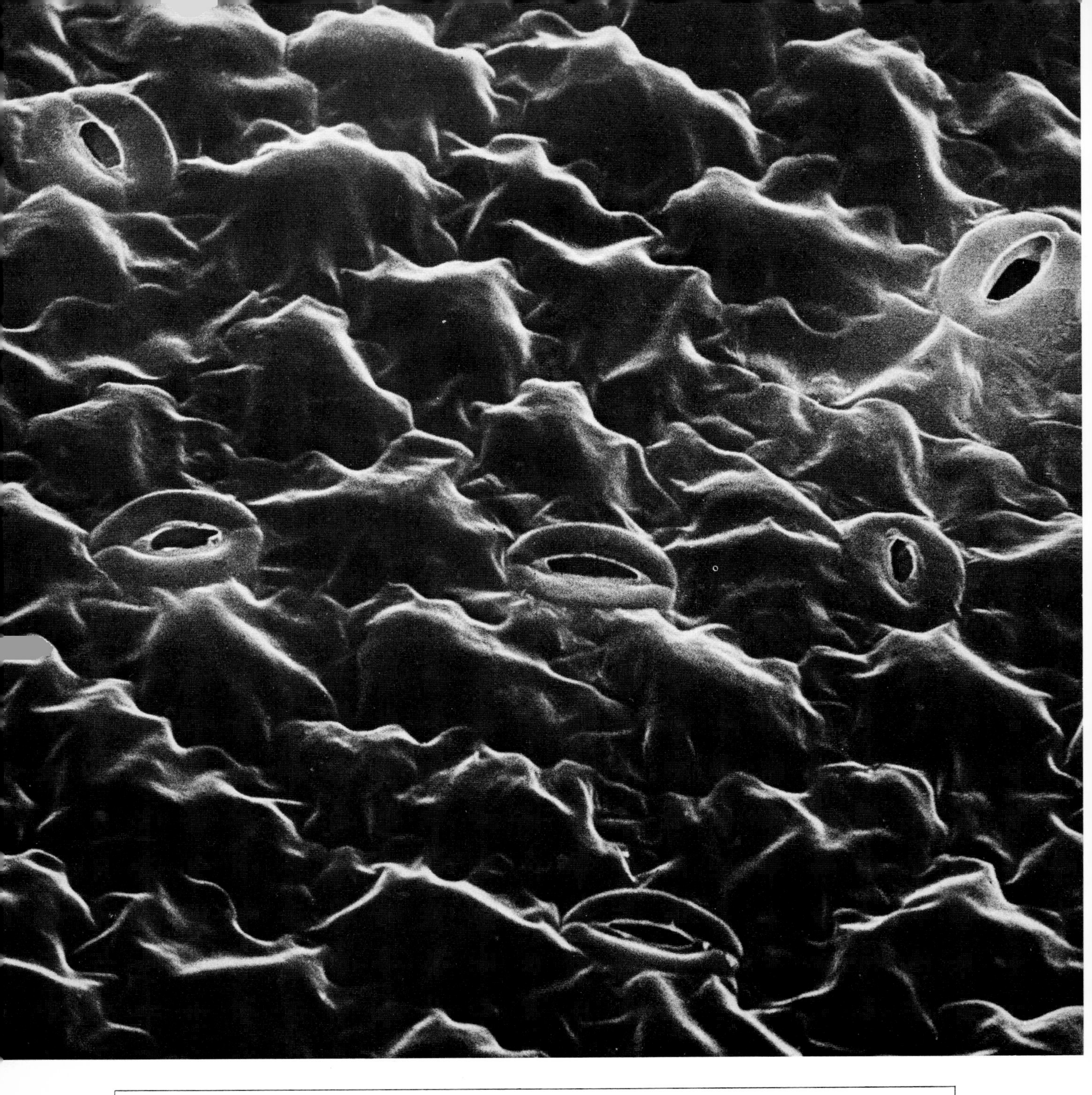

Plate 12 x1400 | Stomata

Stomata on the upper surface of a cucumber *(Cucumis sativa)* **leaf.** The rate of carbon dioxide transfer across the cuticle depends on the difference in CO_2 concentration between the two sides of the cuticle and on the resistance to transfer caused by the epidermis. Almost all the carbon dioxide enters the plant through the stomata so that the resistance caused by the epidermis can be directly related to the number of stomata per unit leaf area and the size of the pore. Both these components are highly variable between and within plant species, as a result of pretreatment, age of the tissue, or due to environmental conditions.

Plate 13 x1400 — Stomata

Stomata on the lower surface of a cucumber leaf. By comparing this photograph with the previous one it can be seen that there are twice as many stomata on the lower compared with the upper surface of the same leaf. A further large variation is in the size of the stomata, both within and between the leaf surfaces. More CO_2 would be expected to enter a cucumber leaf through the lower than the upper leaf surface because of the greater number of stomata, although the stomata on the two sides of the leaf may open or close, to some extent, independently of each other. A feature of dicotyledons such as cucumber is that the stomata are distributed at random over the leaf surface.

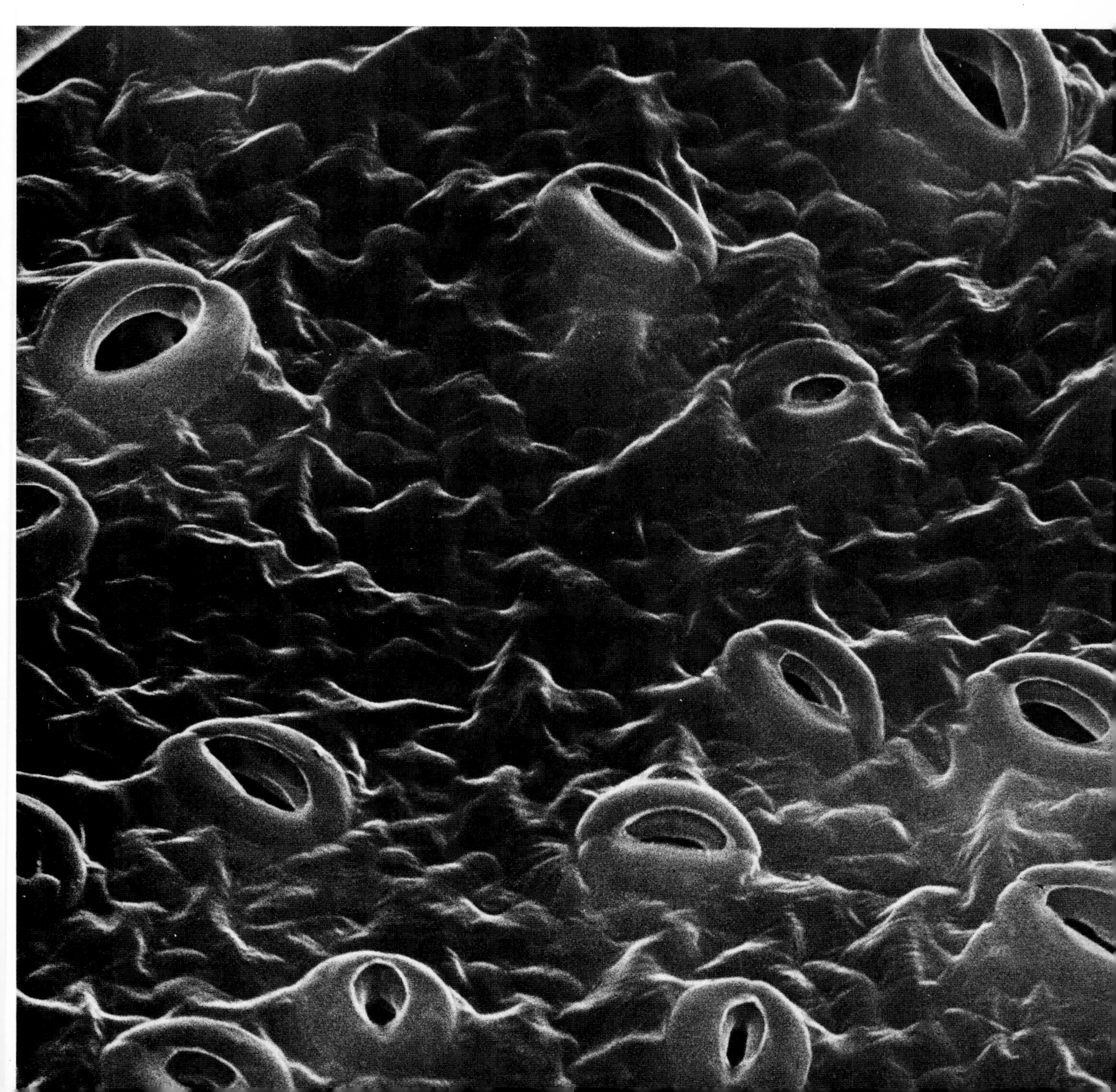

Plate 14 x940 Stomata

Stomata on the lower leaf surface of banana *(Musa sp.).* There are numerous stomata on this leaf surface but even when the stomatal pores are fully open the area of the pores is generally less than 5% of the area of the leaf. The rest of the area of the leaf in banana is coated with wax and is relatively impermeable to CO_2 and water. Banana is a monocotyledon and, as shown in this photograph, the stomata are arranged in parallel rows. In many monocotyledons the rows of stomata follow the veins along the length of the leaf but in banana they run at right angles to the midrib of the leaf.

Plate 15 x2500 — Stomata

Crystals, stoma and epidermal cells of Townsville stylo *(Stylosanthes humilis)*. Townsville stylo has become an economically important forage species in the tropical areas of northern Australia. This photograph illustrates the relationship of a stoma to other cells, including those with crystals embedded in the cutinised walls of some epidermal cells. The stomata are distributed at random among other epidermal cells—characteristic of many dicotyledons.

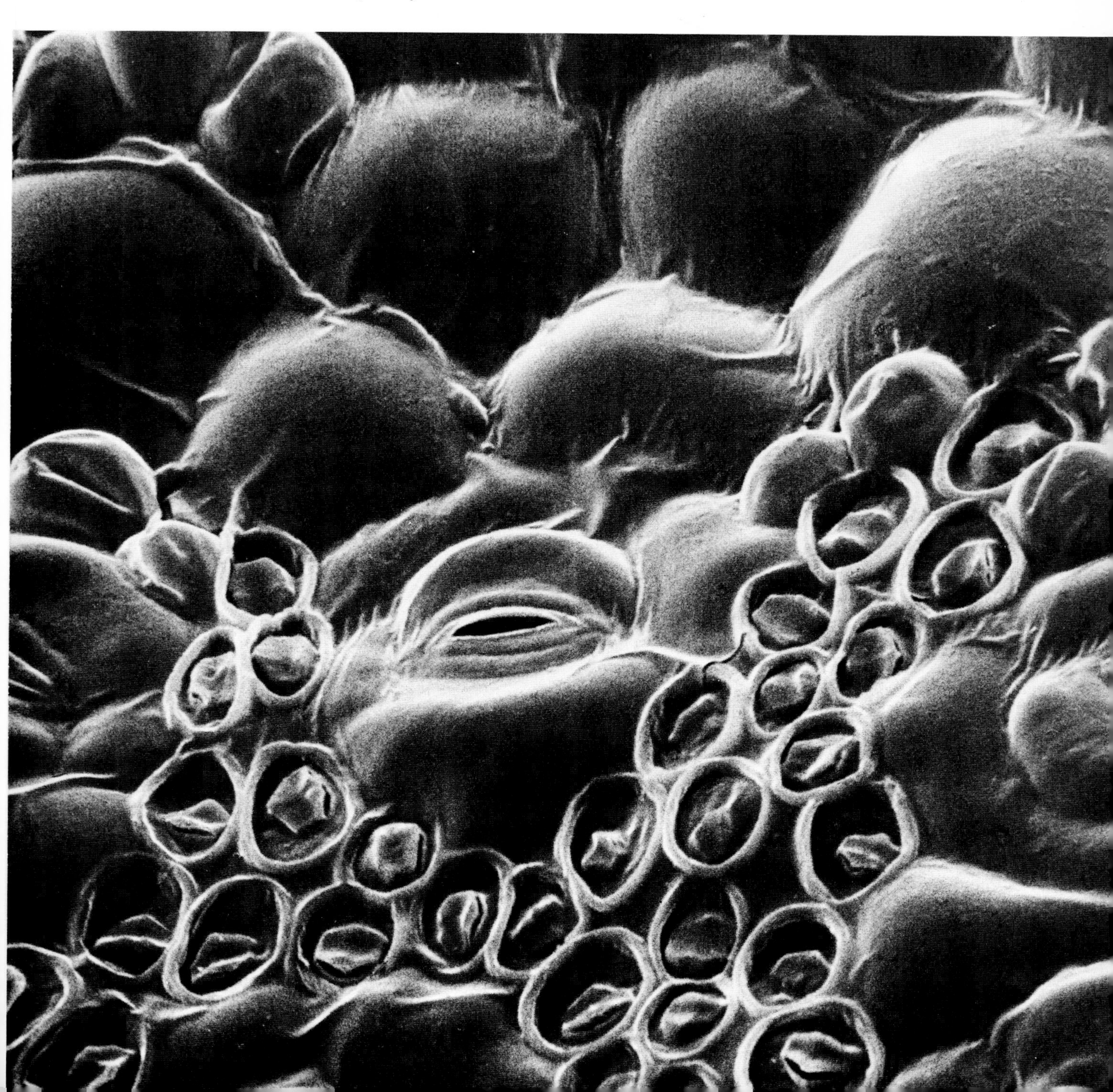

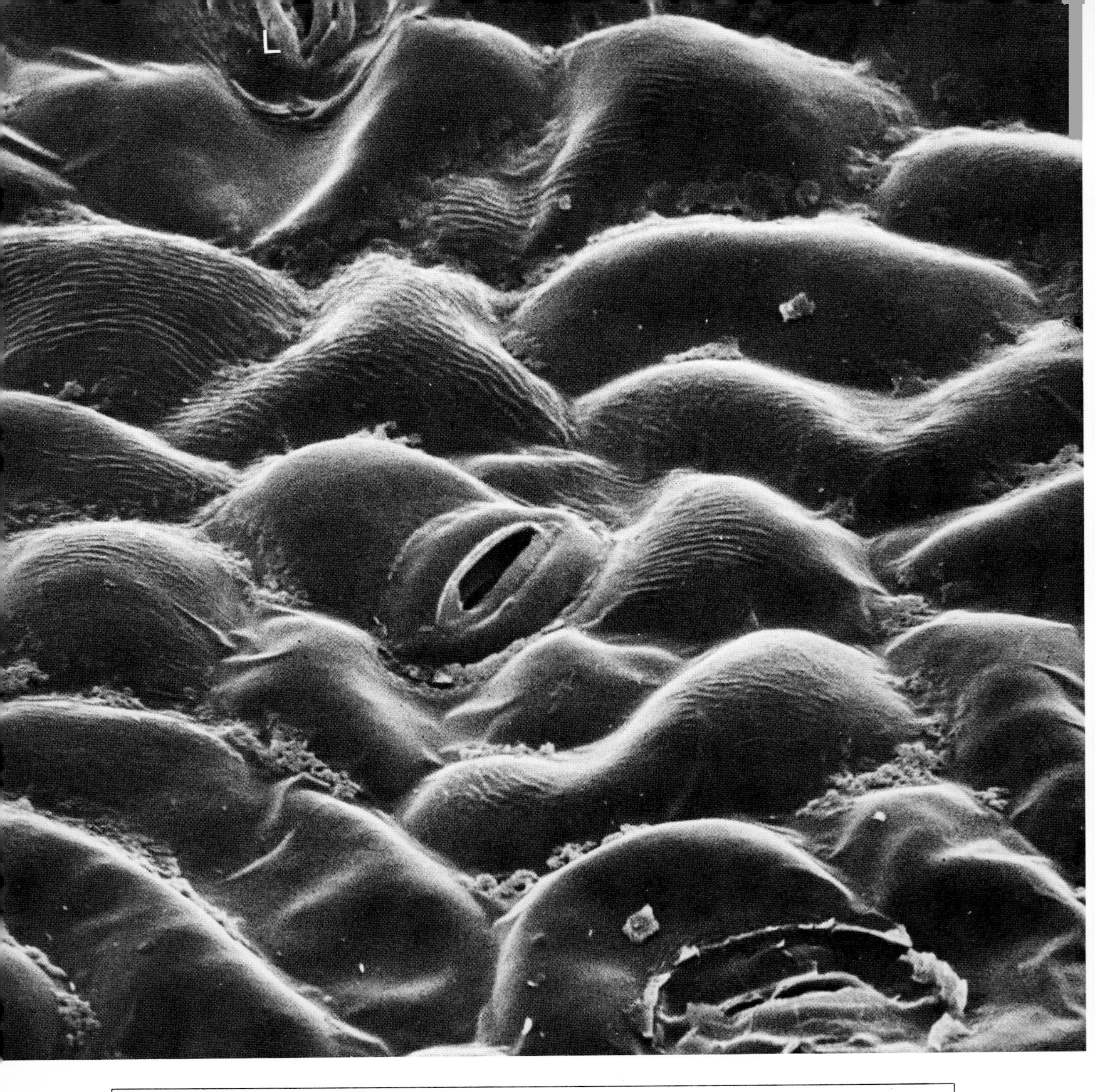

Plate 16 x2100 Stomata

Stomata on the upper leaf surface of cotton *(Gossypium hirsutum).* A feature of the stomata of cotton plants is the pronounced lip (L) at the entrance to the stomatal pore. This is an extension of the cuticle and is common in many species, although the degree of development of this lip is variable. The cuticle in cotton covers the epidermal cells smoothly and it is difficult to locate the boundary between the cells. This leaf was allowed to become slightly water-stressed before being prepared for microscopy and although the stomata have not closed at this stage, other epidermal cells are beginning to shrink and the surfaces are becoming wrinkled.

Plate 17 x2300 — Stomata

Stoma on the upper surface of a leaf of ***Atriplex hastata*****.** This view emphasises the difference in size and shape of the cells making the stomata compared with other epidermal cells. The stomata in this species are at a lower level than other epidermal cells. The cuticle is covered with wax and, in contrast to *Atriplex spongiosa*, there are very few epidermal bladders on any part of the leaf.

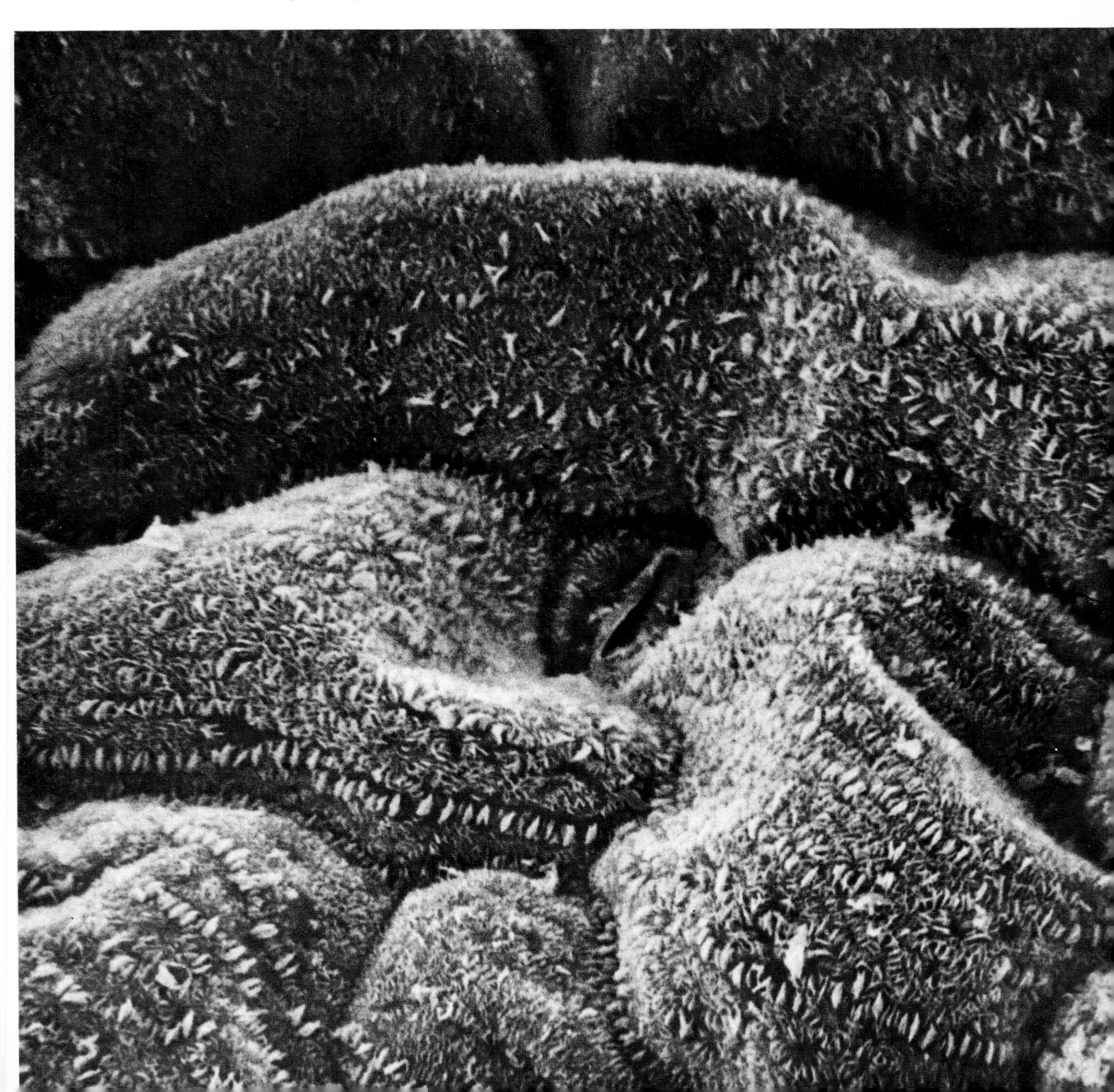

Plate 18 x460 — Stomata

Stomata on the surface of a leaf of *Aloe variegata*. This plant is generally referred to as a succulent because it has developed an extensive system to store water in the leaf or stem. This ability to store water, coupled with a low rate of transpiration, prevents the plant from becoming dehydrated during prolonged dry periods. The low rate of transpiration is achieved by a reduction in the number of stomata on the leaf and by closing the stomata in the daytime and opening them only at night. This latter feature allows the plant to partly overcome the high potential evaporation caused by the high radiation environment in the daytime. This feature is a disadvantage to carbon dioxide uptake which normally takes place in the light. However this type of plant has developed a system of taking up CO_2 at night and transforming it into other compounds in the daytime.

Plate 19 x7900 Stomata

A stoma in the epidermis of a cucumber leaf. This stoma is characteristic of stomata in dicotyledons because there are the two clearly defined guard cells surrounding the pore. The guard cells are fully extended in this example because the plant had been in the light before being prepared for the Scanning Electron Microscope. Furthermore, the leaf was cut off the plant which would almost immediately cause a reduction in turgor in the epidermal cells. This, in turn, would tend to pull the guard cells apart. If the leaf had been left in this condition the guard cells would eventually collapse because of water stress, and the pore would close.

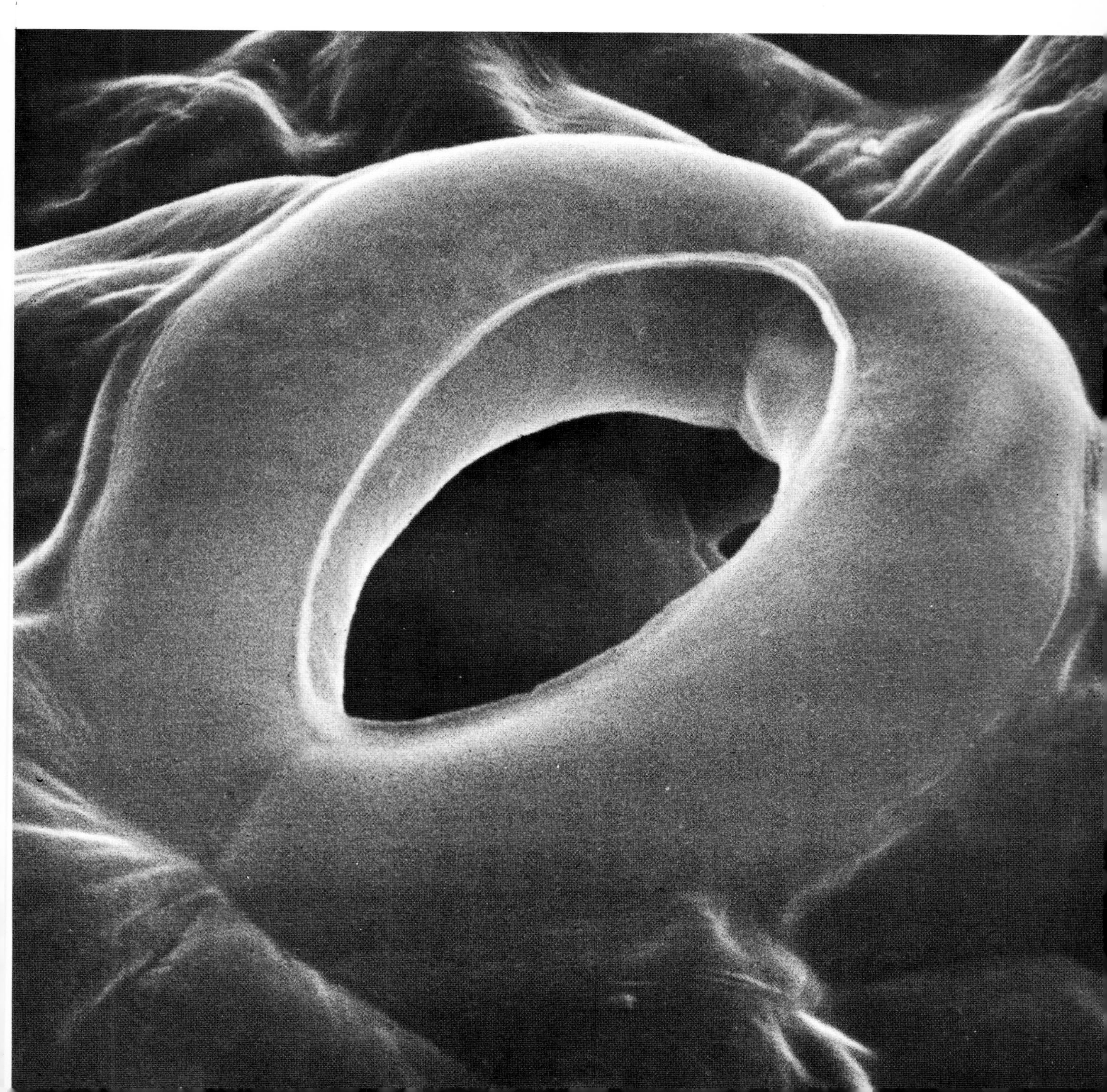

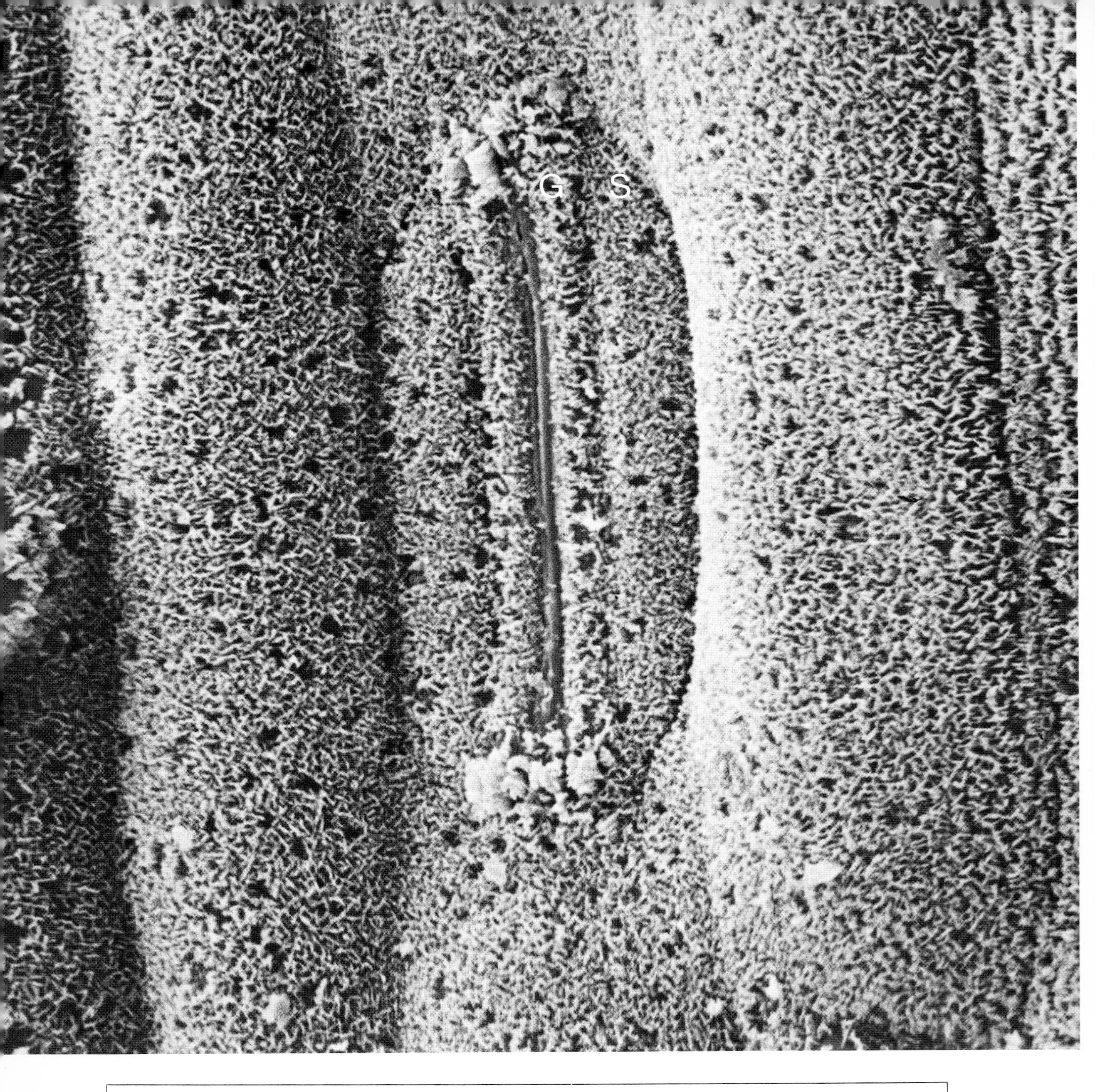

Plate 20 x2300 | Stomata

A stoma in the upper surface of a wheat leaf *(Triticum aestivum,* variety Raven). Wheat is a monocotyledon and the outline of the guard cell (G) and the subsidiary cell (S) in the stomatal complex can be seen. The stoma is closed because the leaf was water stressed before being prepared for the Scanning Electron Microscope. The combined effect of wax and the closed stoma prevents water from escaping out of the leaf.

Plate 21 x4400 | Stomata

A stoma on the lower surface of a wheat leaf (variety Raven). Wax on this leaf surface appears as long rodlets lying on the cuticle. The stoma in wheat, when open, appears as a long slit which is in contrast to the oval appearance of the stomatal pore in cucumber.

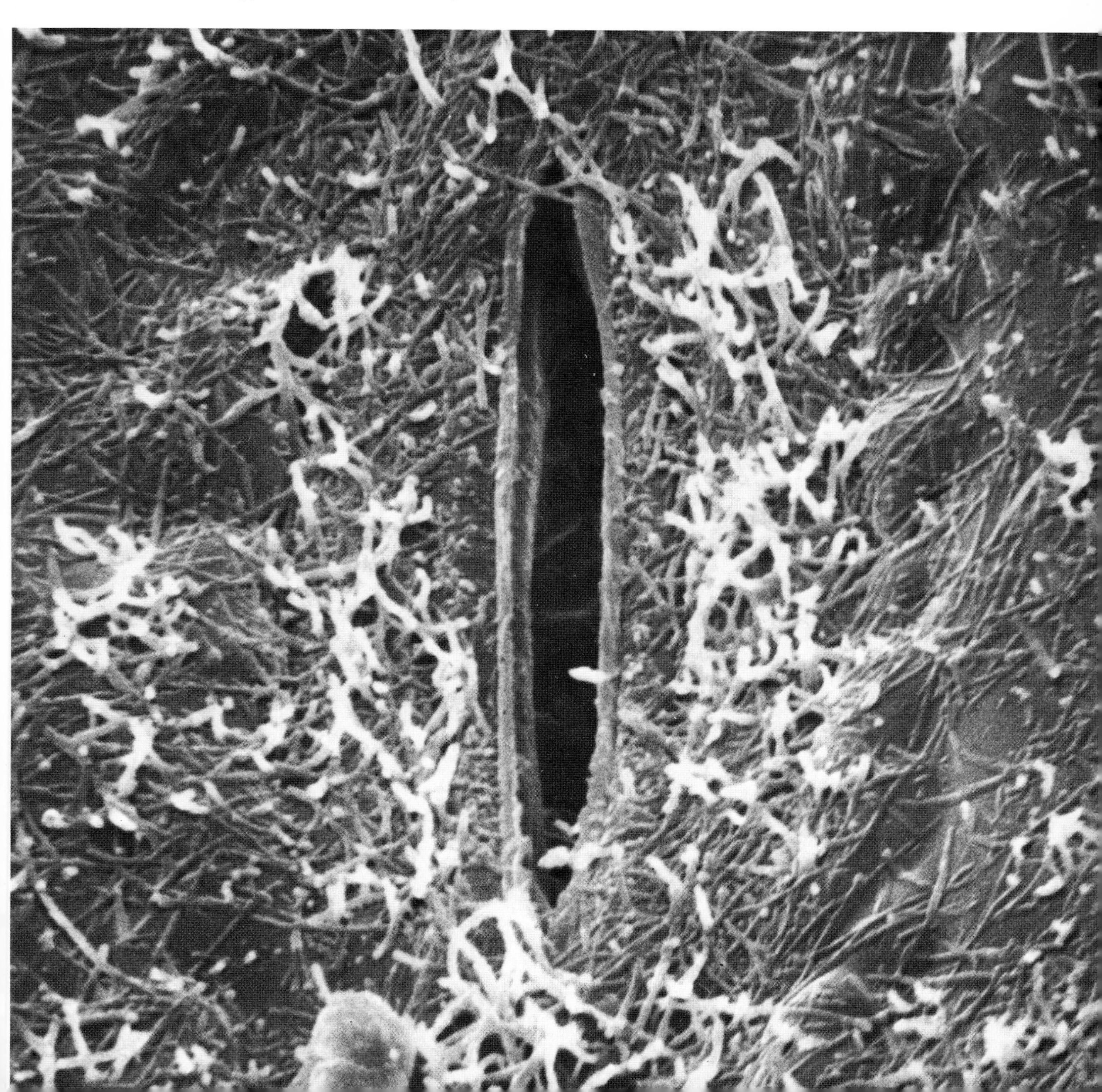

Plate 22 x4100 **Stomata**

A stoma on the leaf of maize. This is another example of a stoma in a monocotyledon which has the guard cell and a subsidiary cell. Although wax plates cover most of the surface they are apparently very much reduced on the guard cell surface.

Plate 23 x5600	Stomata

A stoma on a leaf of banana. In this species there is a pronounced cuticular lip over the stoma although the guard cells can still be clearly seen within the pore.

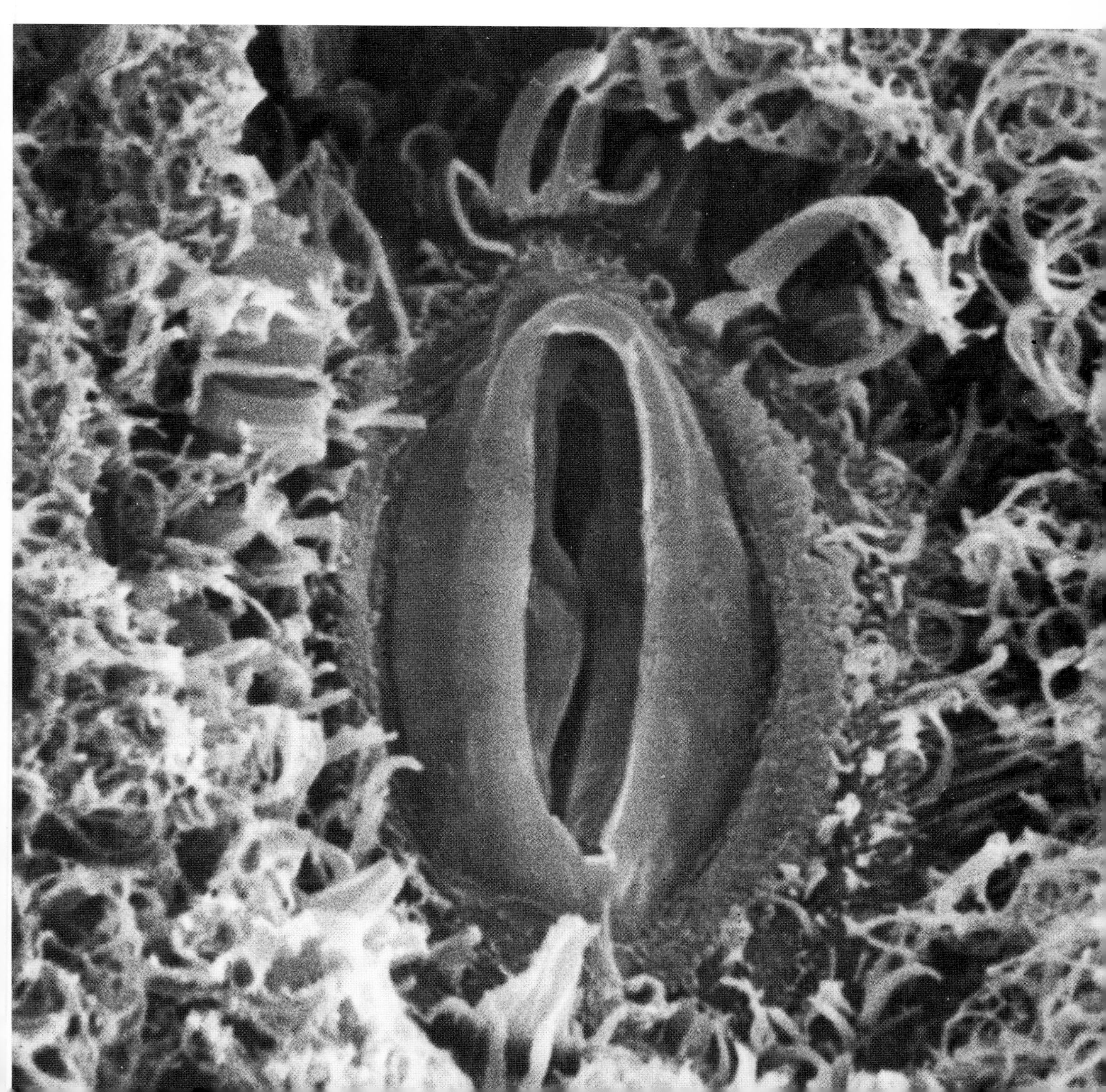

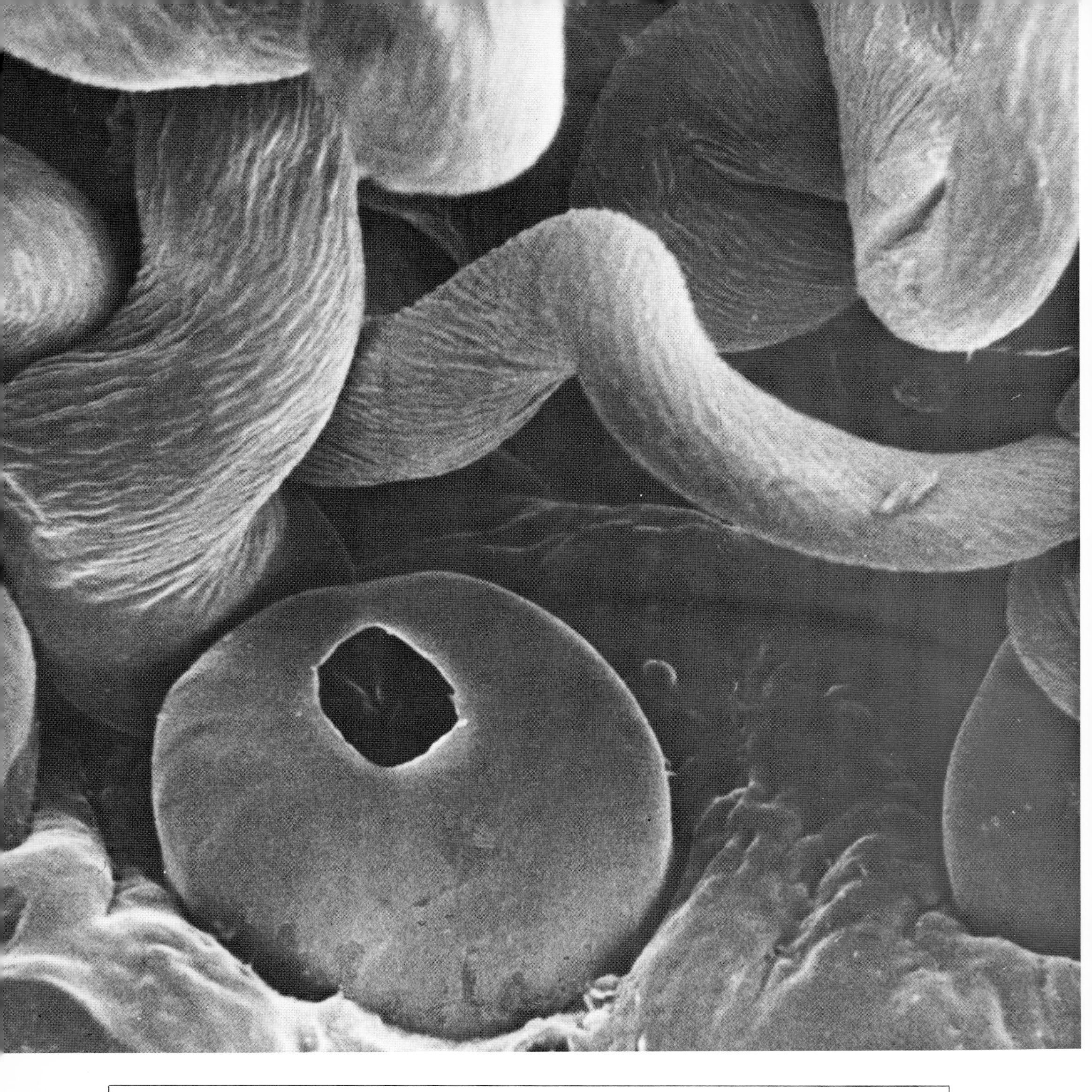

Plate 24 x6000 Stomata

A stoma on the leaf of the New Zealand Christmas Tree. The cuticular lip is more extensive than in banana and causes the stoma to be raised above the surface of the epidermal cells. The fibrous like nature of the hairs on the leaf can also be seen in this photograph.

Plate 25 x3900	Stomata

Stomata on a pine tree *(Pinus sp.)* **needle.** The stomata on pine trees are normally sunken below the surface of the needle. This is regarded as a xeromorphic feature i.e. an adaption for conserving water. Pine has several other features in this category and these include the needle-like leaf, close packed cells in the leaf and a thick cuticle.

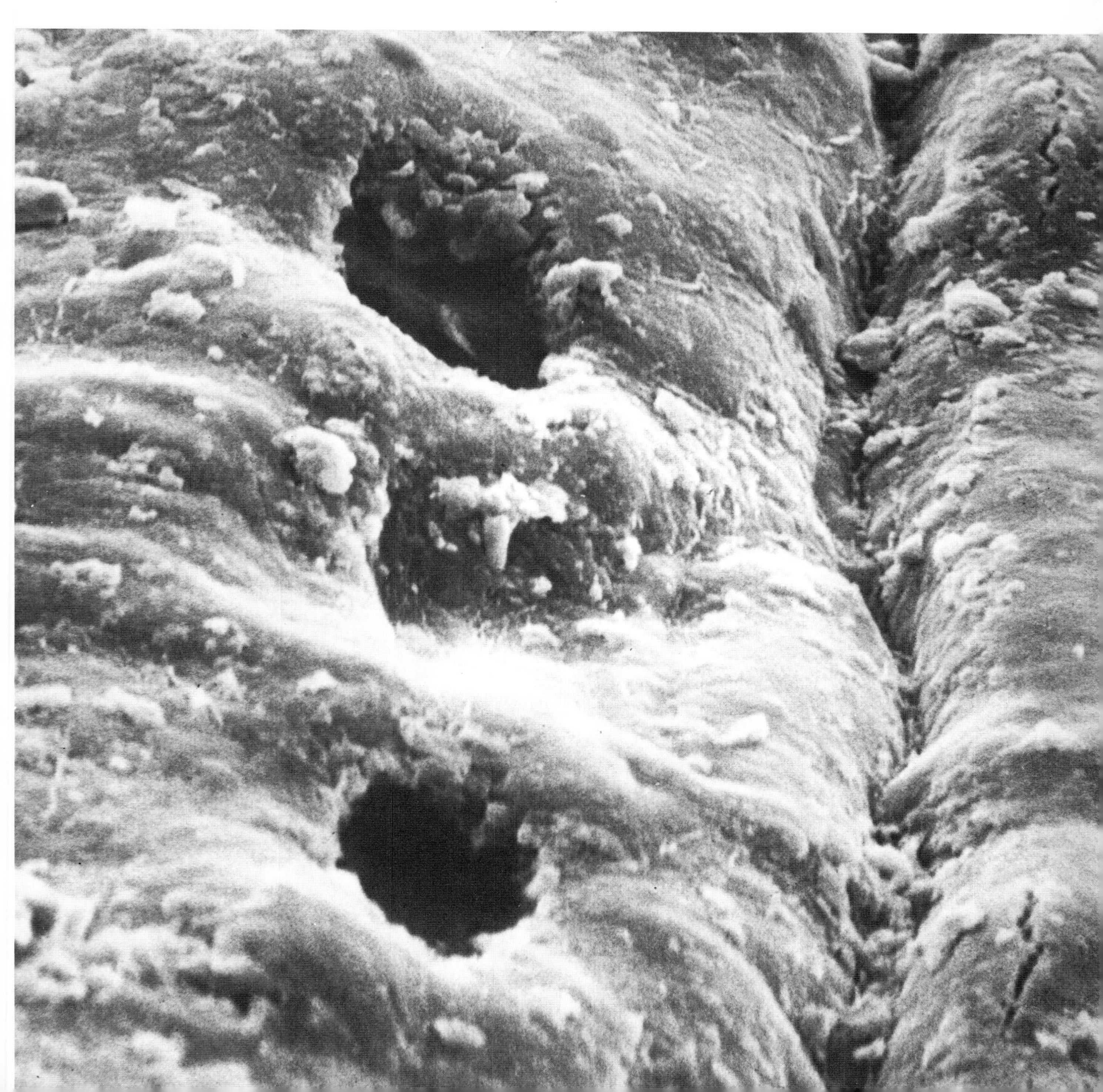

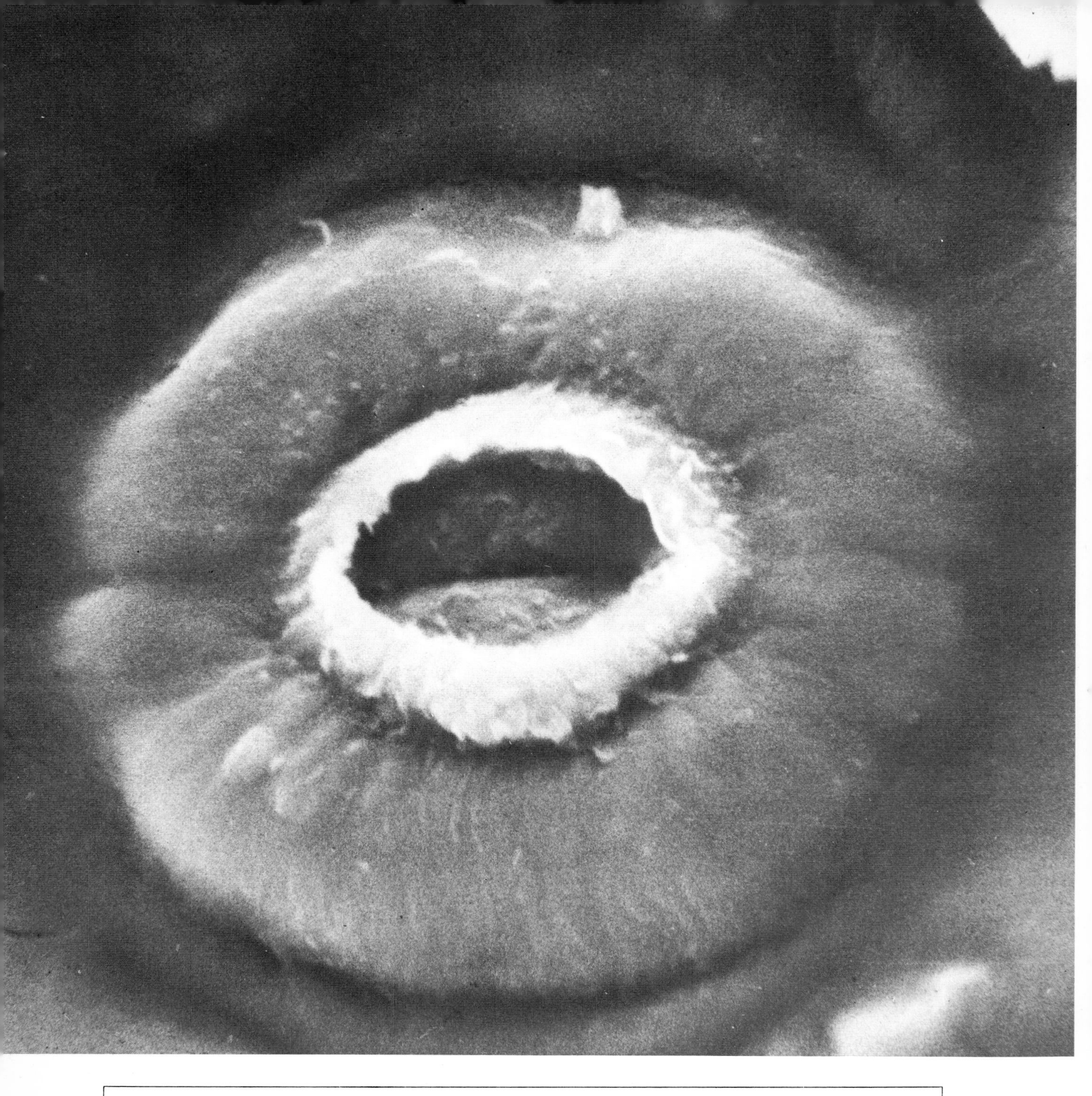

Plate 26 x7200 — Stomata

A stoma on the upper leaf surface of pineapple *(Ananas comosus).* The stomata on pineapple are small and occur in rows in the troughs on the leaf. In this position they occur under the scales on the leaf. These stomata have subsidiary cells but they are often located under the guard cells and not alongside them. In some species in the pineapple family, the subsidiary cells meet under the guard cells and provide an extra structure in the pathway of CO_2 into the leaf.

Plate 27 x6300 Stomata

A stoma on the upper leaf surface of tomato *(Lycopersicon esculentum).* The guard cells are raised above the level of the other epidermal cells.

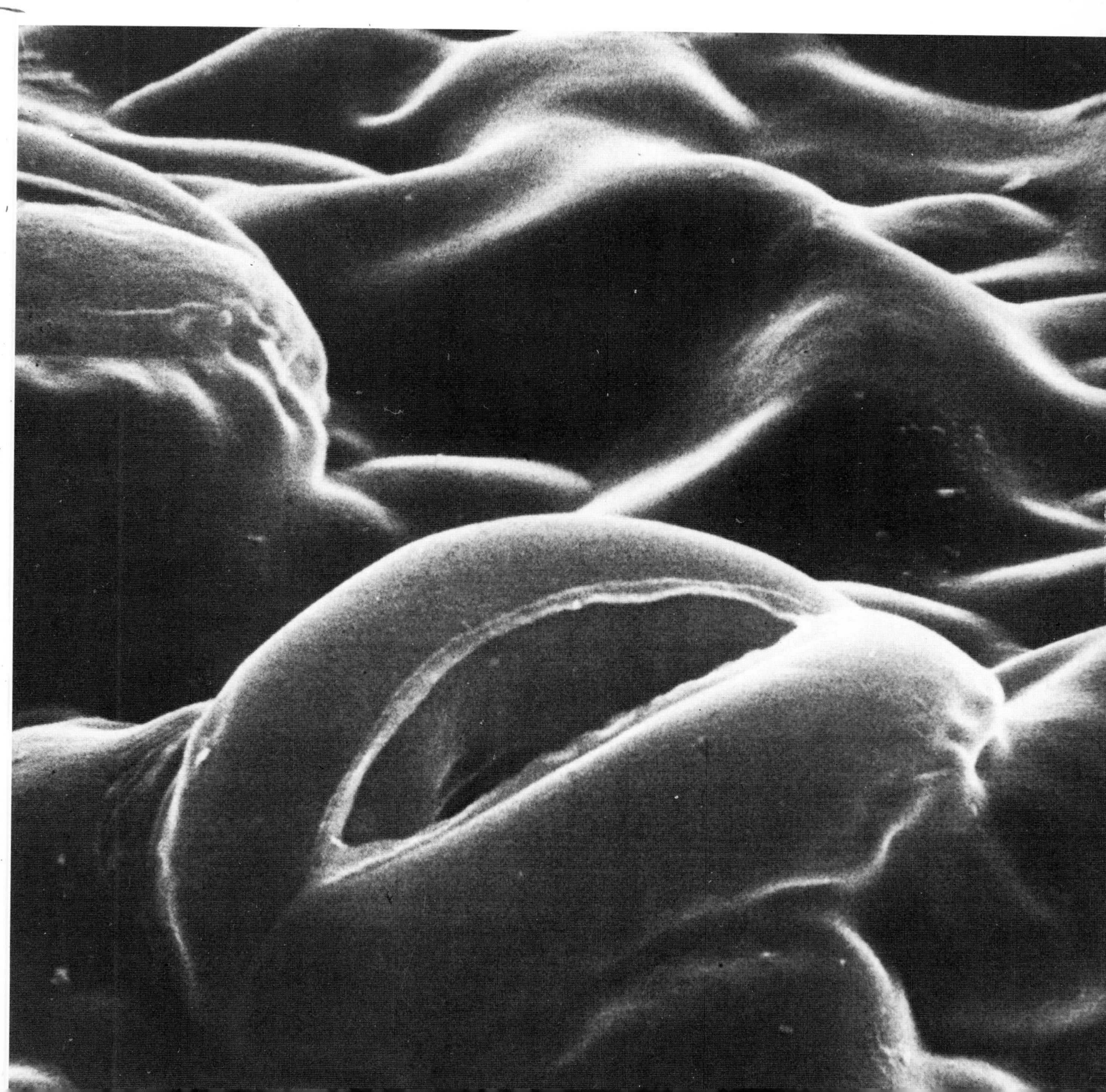

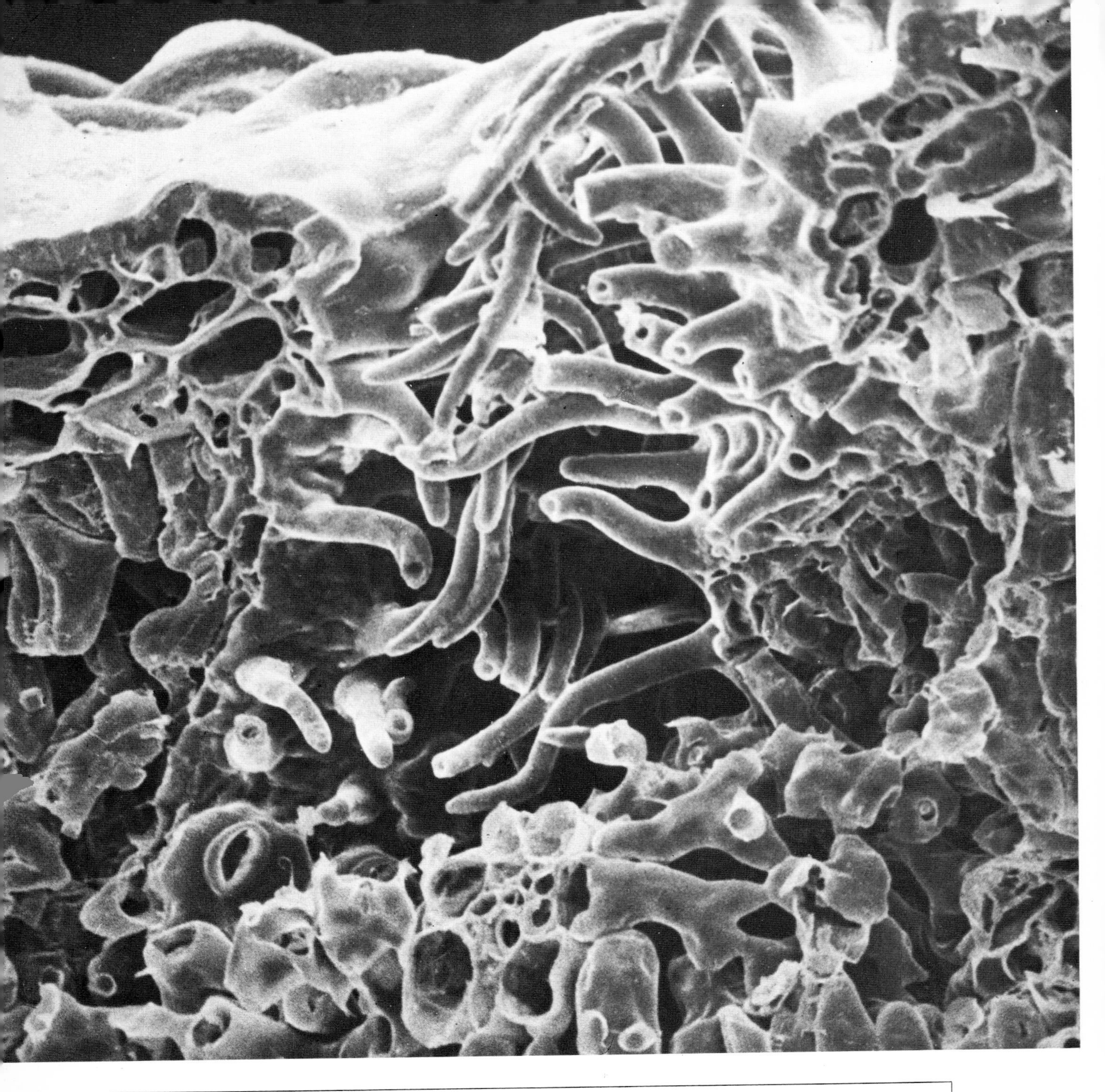

Plate 28 x1000 **Stomata**

Transverse view of the pits with stomata on the lower surface of a leaf of oleander *(Nerium oleander).* The upper leaf surface of oleander has few stomata and occasional hairs are randomly distributed over the surface. The lower leaf surface has clumps of hairs which arise from invaginations of the cuticle. Numerous stomata are scattered at the base of each pit and below the hairs around the epidermal cells, which are overlaid with a thick cuticle. This leaf therefore has numerous anatomical features characteristic of plants in xerophytic situations.

Plate 29 x7900	Inside the Leaf

The mesophyll cells, as seen through a stoma, on the lower leaf surface of cucumber. The mesophyll cells within the leaf are important because they are the site from which water vapour evaporates and carbon dioxide goes into solution. By looking through a stoma it is possible to see several mesophyll cells, which in this example are the spongy mesophyll cells because the stoma is on the lower side of the leaf.

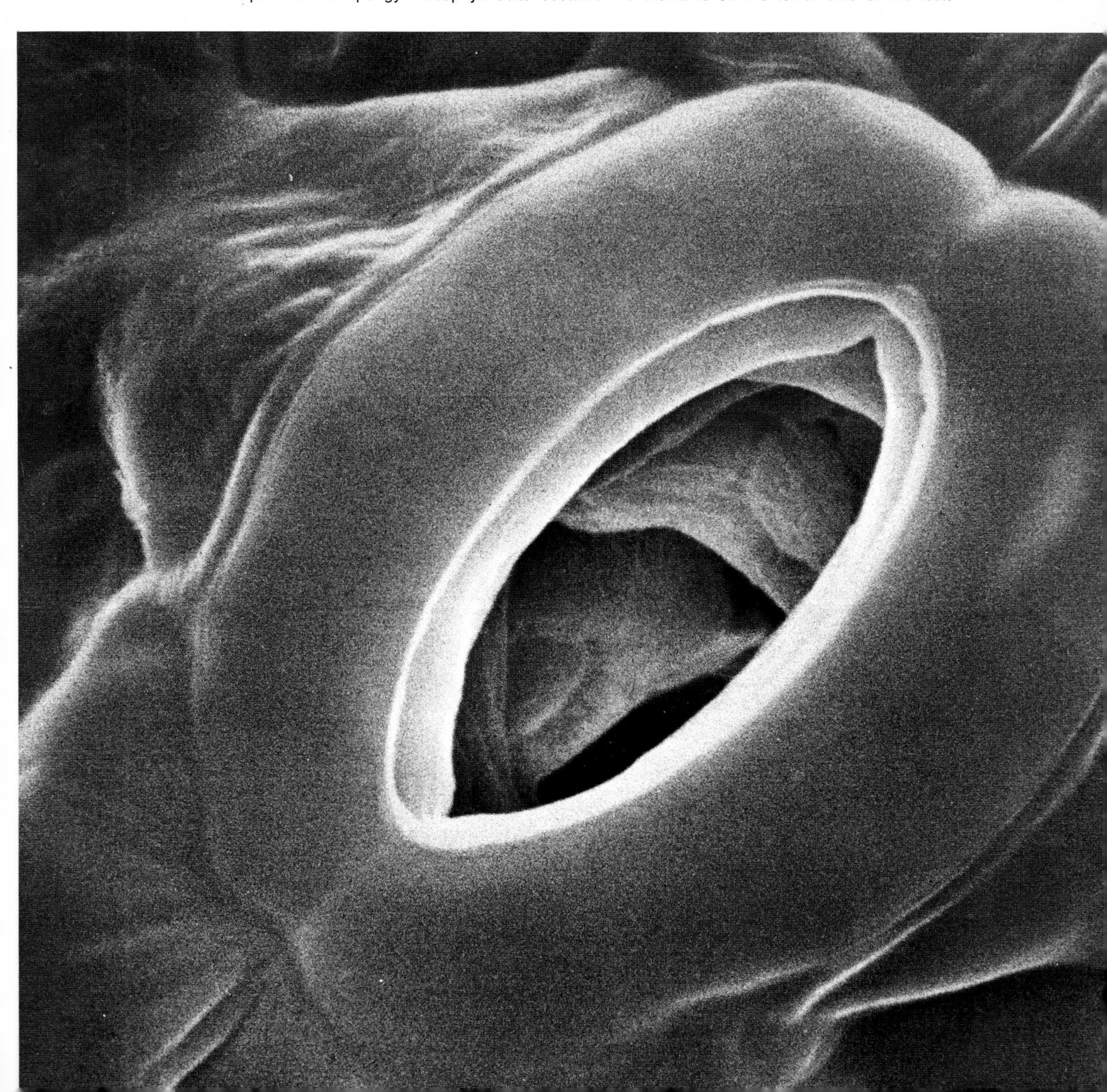

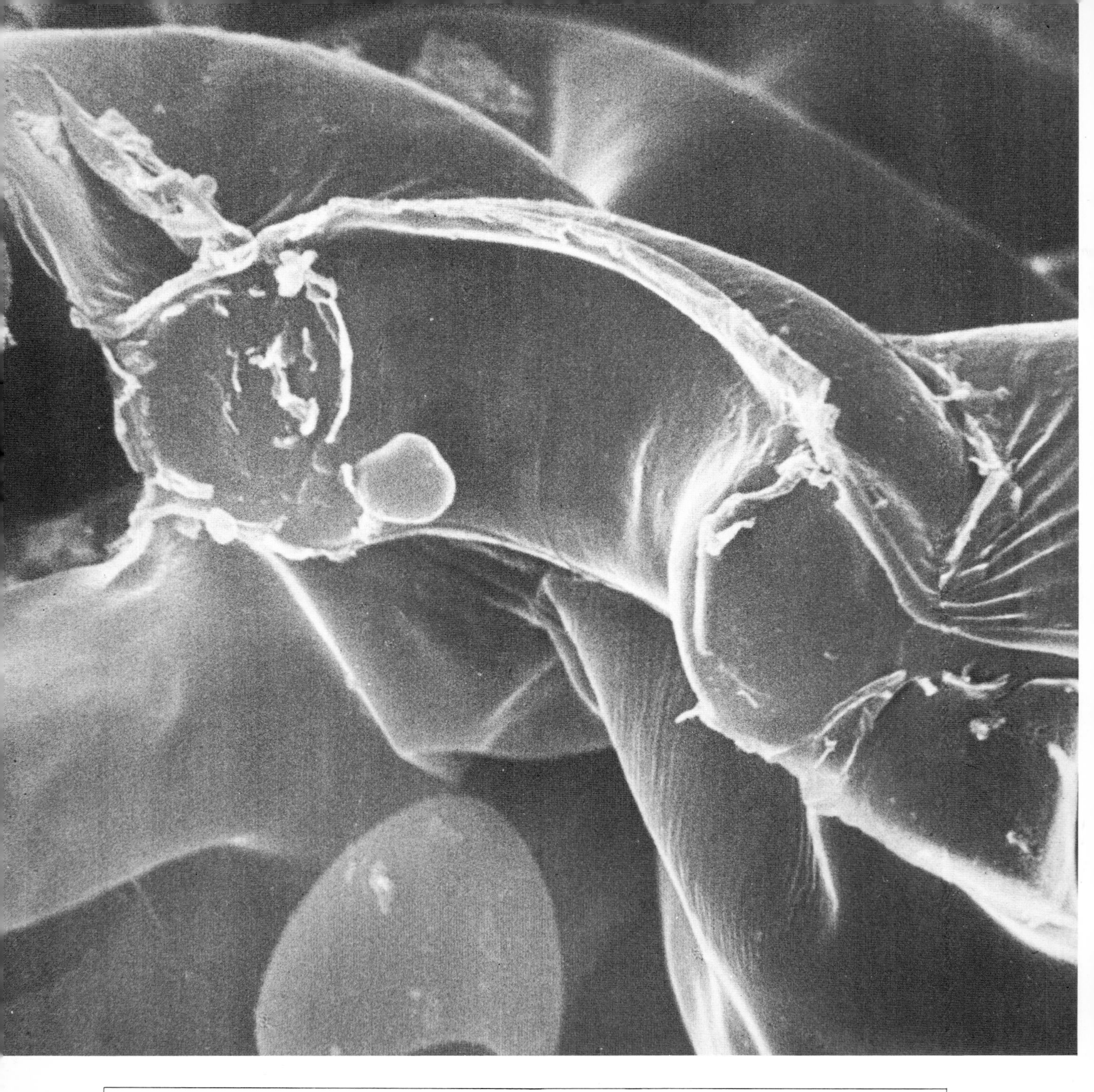

Plate 30 x4300 — Inside the Leaf

Single guard cell of a stoma in the broad bean *(Vicia faba).* The plant material was fractured in liquid nitrogen and a stoma was broken in half, revealing the shape of a single guard cell. There is a small cuticular lip which is continuous with the cuticle overlying the epidermal cells. Just below the guard cell the leaf opens out and that space is the substomatal cavity.

Plate 31 x6000	Inside the Leaf

Transverse view of a New Zealand Christmas Tree stoma. The feature of stomata on the New Zealand Christmas Tree is the extensive development of the cuticular lip over the stoma. The guard cells occur at the level of the epidermis and open and close to regulate the exchange of CO_2 and water vapour. However, the air entering the leaf has first to pass through a pore of fixed dimensions, and therefore fixed resistance, before entering the ante-chamber above the guard cells. This rigid structure, built over the stoma, is an extension of the cuticle and consists primarily of lipids of high molecular weight.

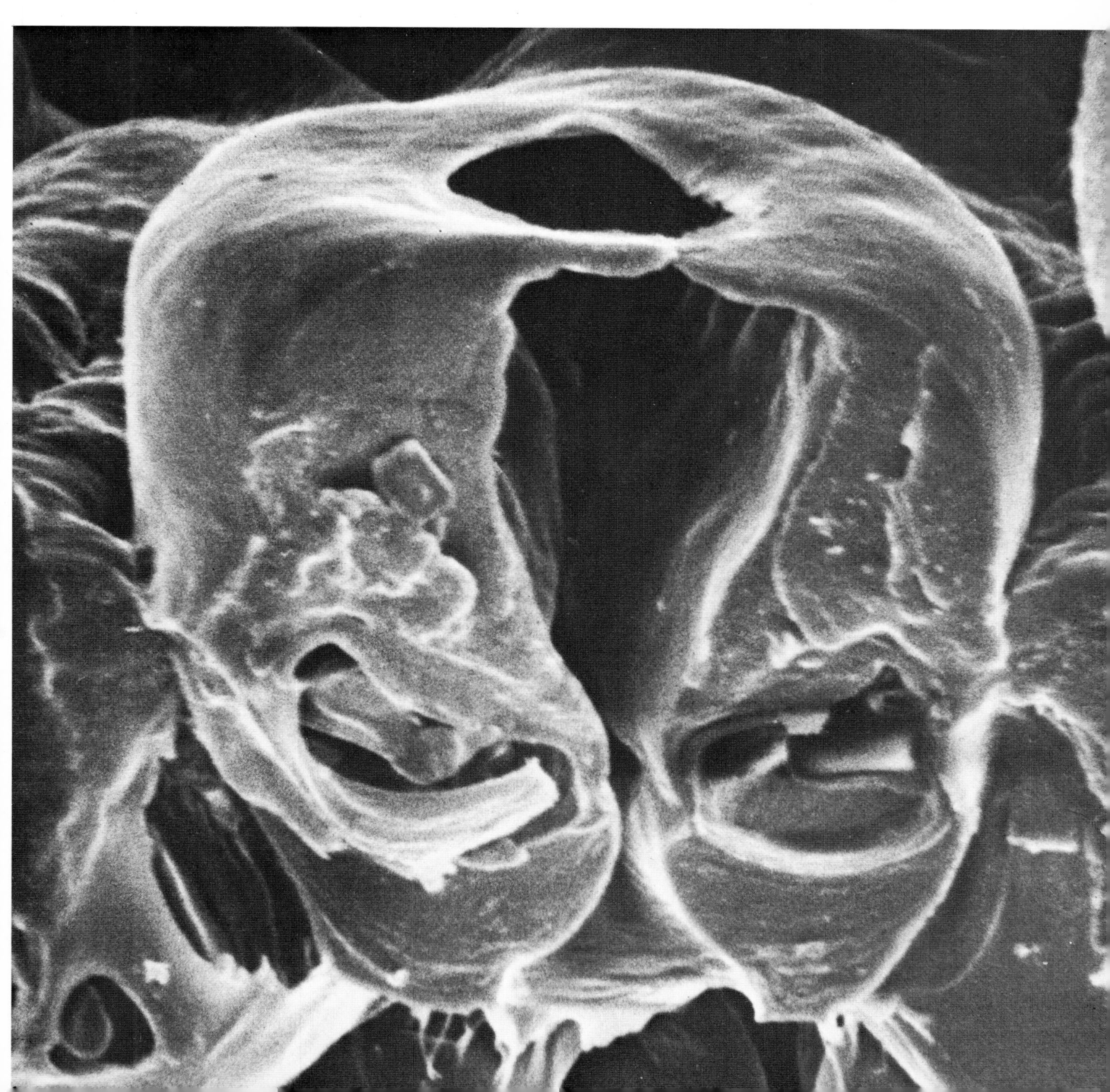

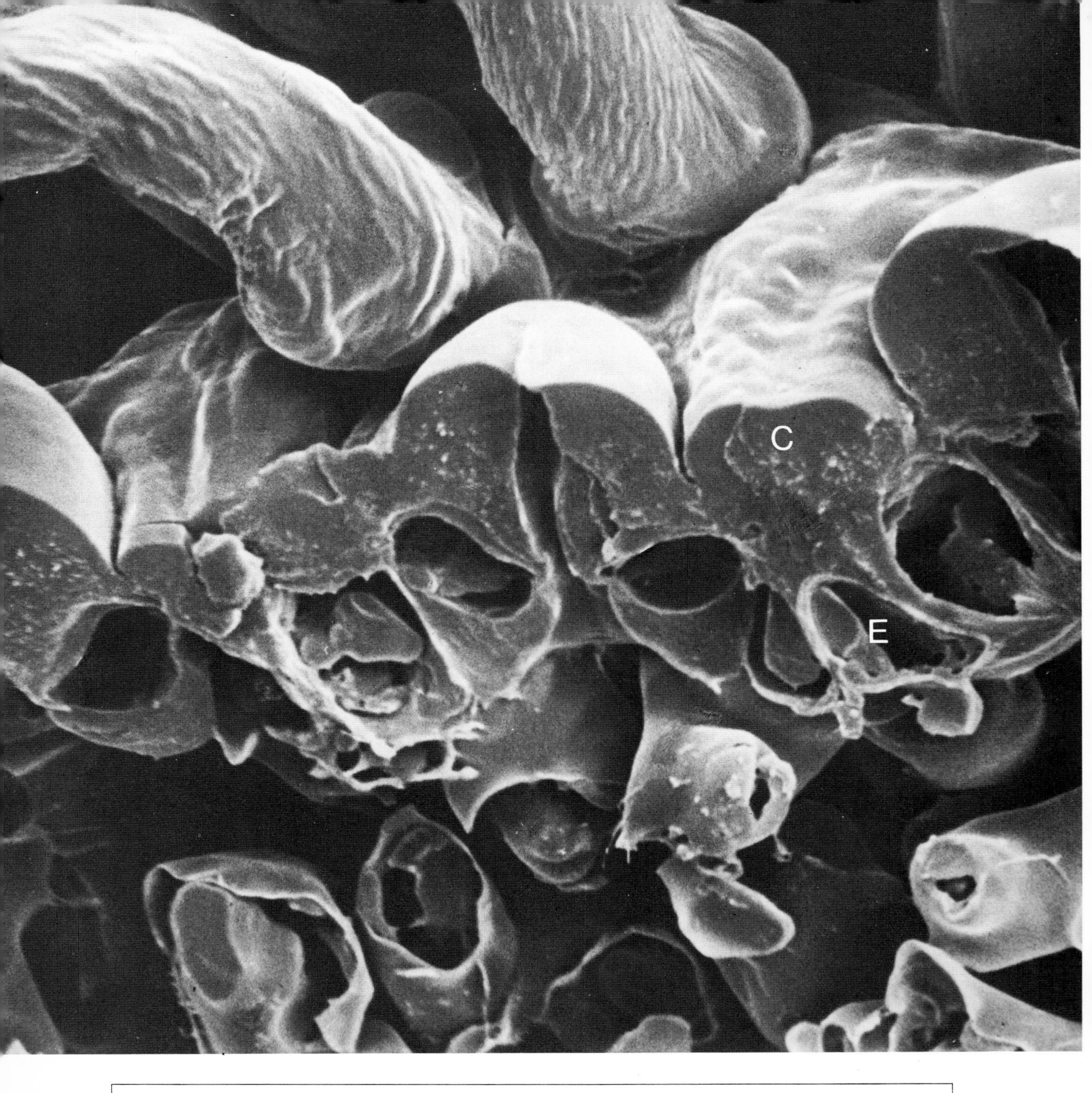

Plate 32 x3100 — Inside the Leaf

The cuticle on the lower surface of a leaf of the New Zealand Christmas Tree. This cross-section through the leaf emphasises the thick, continuous nature of the cuticle (C) overlying the epidermal cells (E). The cuticle very clearly defines the outer surface of the plant and separates the air outside the leaf from that inside the leaf.

Plate 33 x1000 — Inside the Leaf

Transverse view through a stoma and substomatal cavity of *Gomphrena globosa.* The cuticle in this leaf is a thin layer over the outer surface of the large epidermal cells (E). The volume of the guard cell is very small compared with the epidermal cells, which are unspecialised and may act as a storage system for water. It is necessary for the guard cells to be well supplied with water and the pathway for the transport of water is through the cell walls from the xylem elements to the stoma. In this example there are three cells between the xylem and stoma, the bundle sheath cell (B), mesophyll cell (M) and epidermal cell (E). The carbon dioxide passing from outside to inside the leaf is restricted as it passes through the stomata but, as shown here, the substomatal cavity is large and opens into the intercellular passages of the leaf.

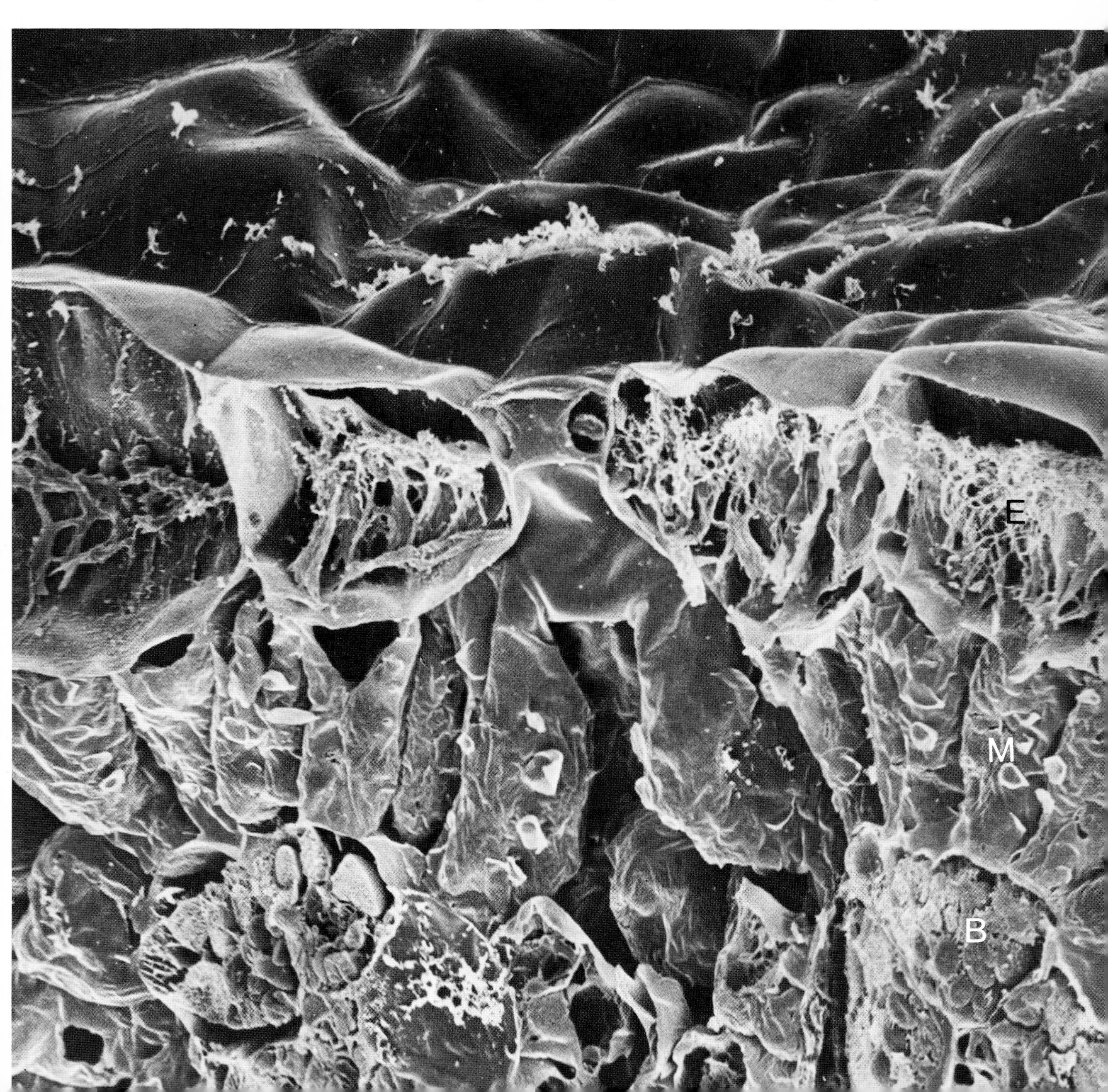

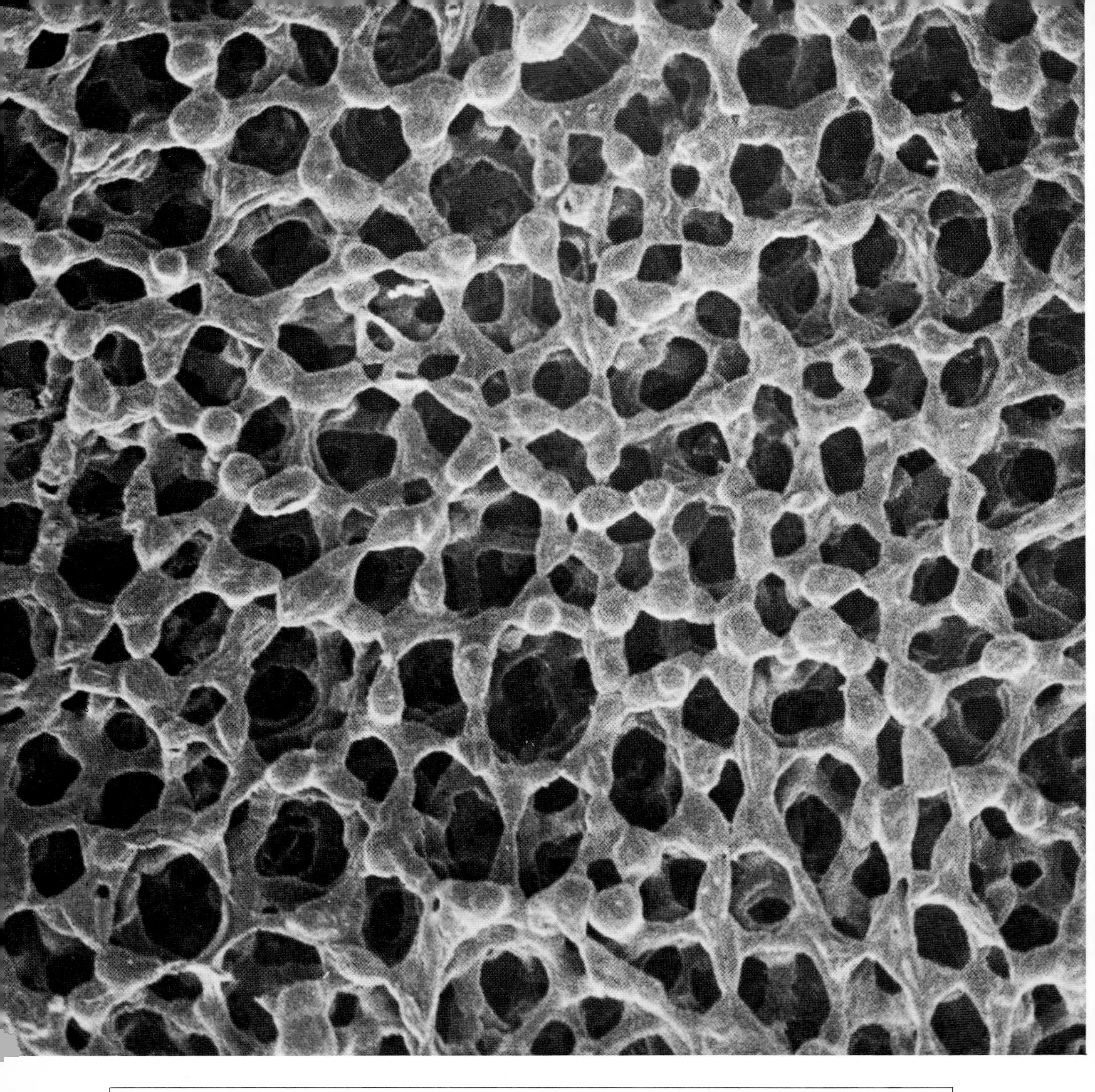

Plate 34 x290 — Inside the Leaf

Spongy mesophyll cells of a broad bean leaf. The organisation of the cells inside the leaf can be seen more easily if the epidermis is removed, and so in this example the lower epidermis was stripped off to show the extensive development of the intercellular air spaces among the spongy mesophyll cells. Some stretching of the cells occurred during the preparation of the specimen. About eight stomata would be present in an area of the leaf shown in this example. The porous nature of the cells in the leaf will allow freedom for the diffusion of CO_2 and H_2O vapour within the leaf.

Plate 35 x520 | Inside the Leaf

Mesophyll cells in a leaf of *Gomphrena globosa. Gomphrena* is a C_4 type plant and the arrangement of cells within the leaf is different to the bean or other C_3 type plants. However the appearance of the cells within the leaf when the epidermis has been stripped off is not dissimilar from bean. There are large air gaps which ramify throughout the leaf. The bundle sheath arrangement of cells, which will be evident in later photographs, does not appear here.

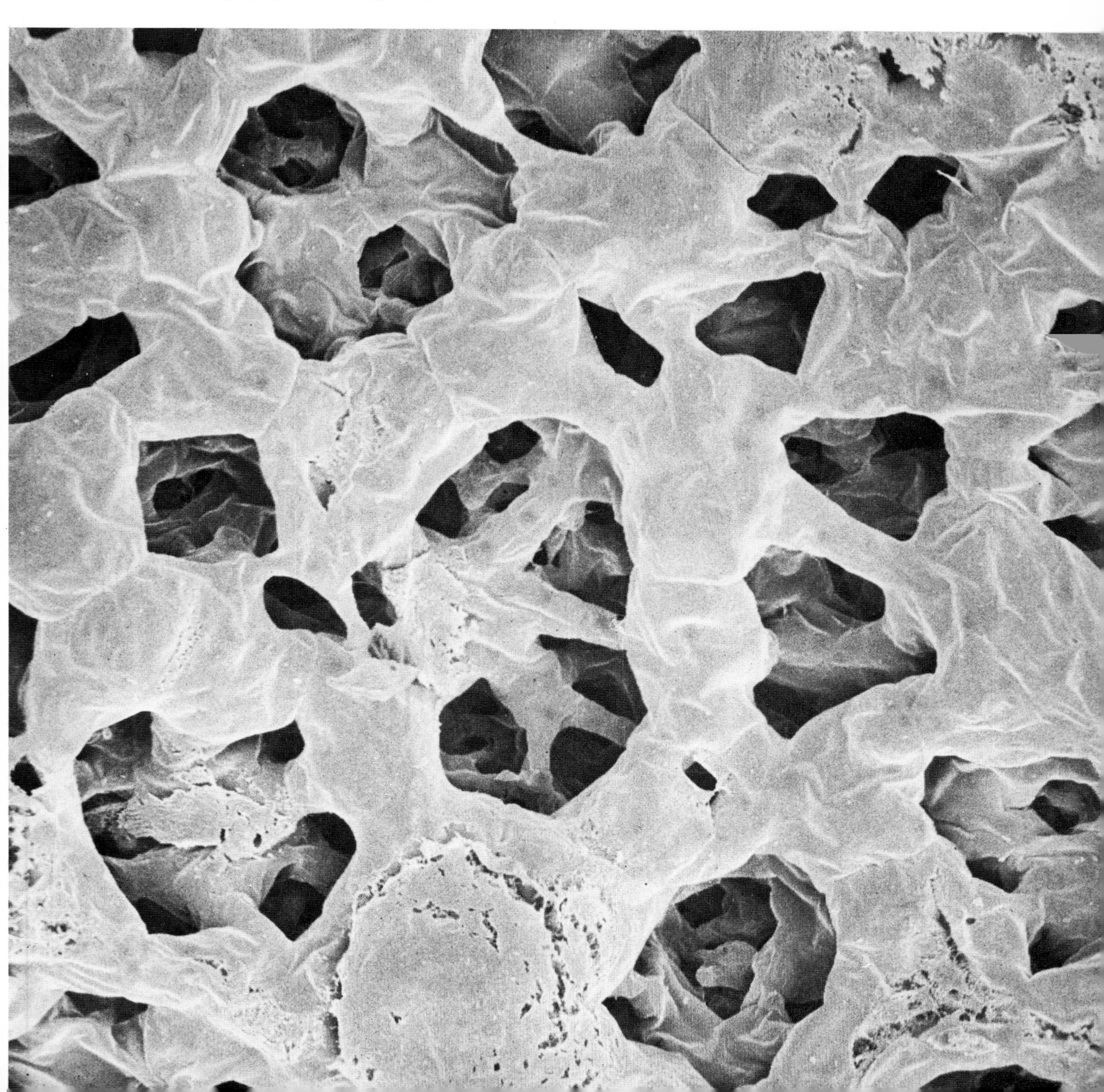

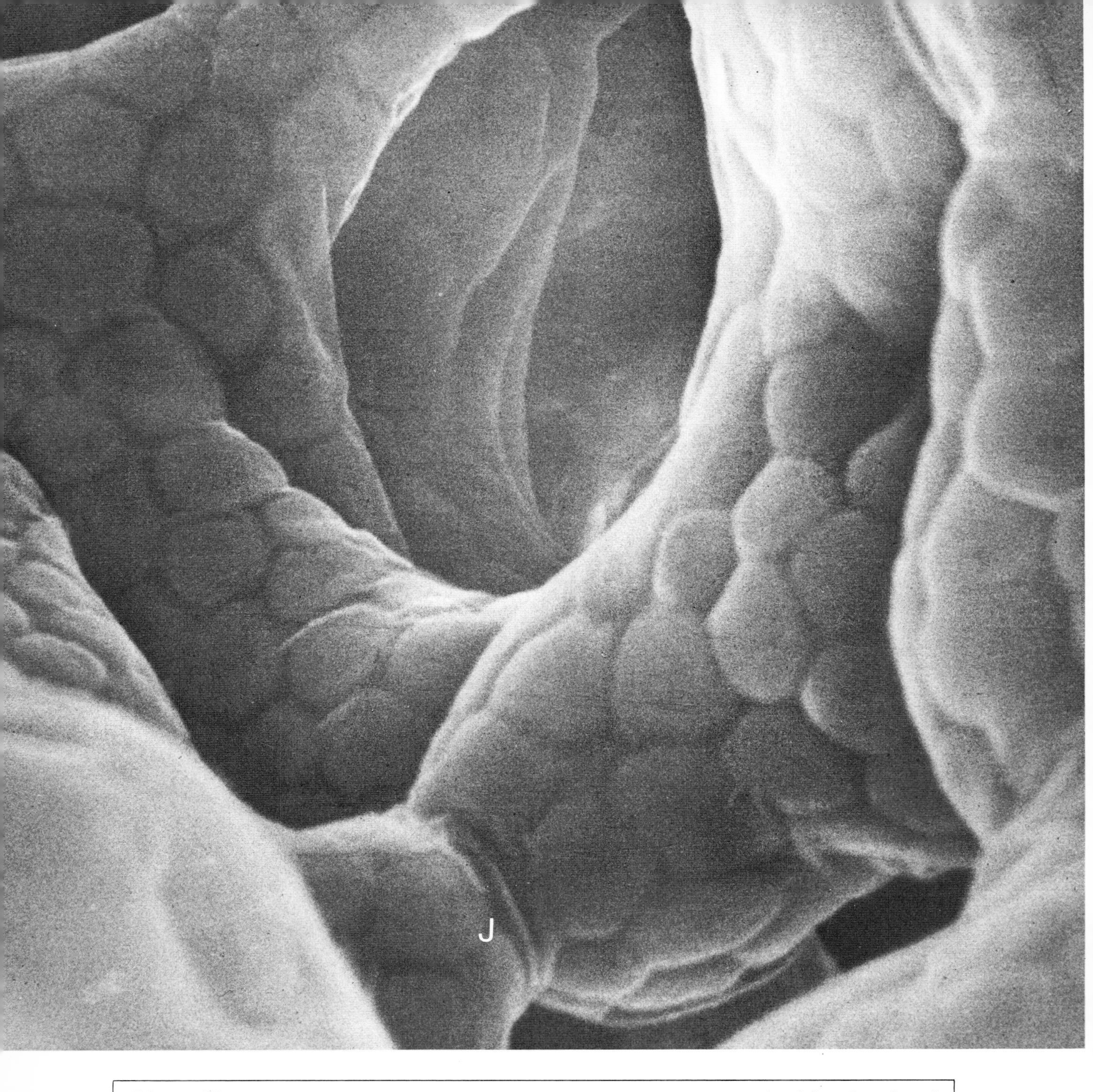

Plate 36 x4200 | Inside the Leaf

Spongy mesophyll cell walls in a broad bean leaf. The honeycomb pattern on the cell wall is due to an impression made on the cell wall by chloroplasts within the cell. Normally the cell walls are wet and when carbon dioxide from the outside air reaches the cell, it enters into solution on these walls. This is the first step of many cellular reactions in the pathway of carbon into the plant. The junction (J) between two cells is also shown in this photograph.

Plate 37 x3400 — Inside the Leaf

Palisade cells in a tomato leaf. The upper epidermis has been stripped from the leaf to reveal the palisade cell layer. This layer is tightly packed together, except in the region of the substomatal cavity, and it would be difficult for gases to diffuse among the cells. The territory occupied by the chloroplasts can also be seen by the impressions made on the cell wall.

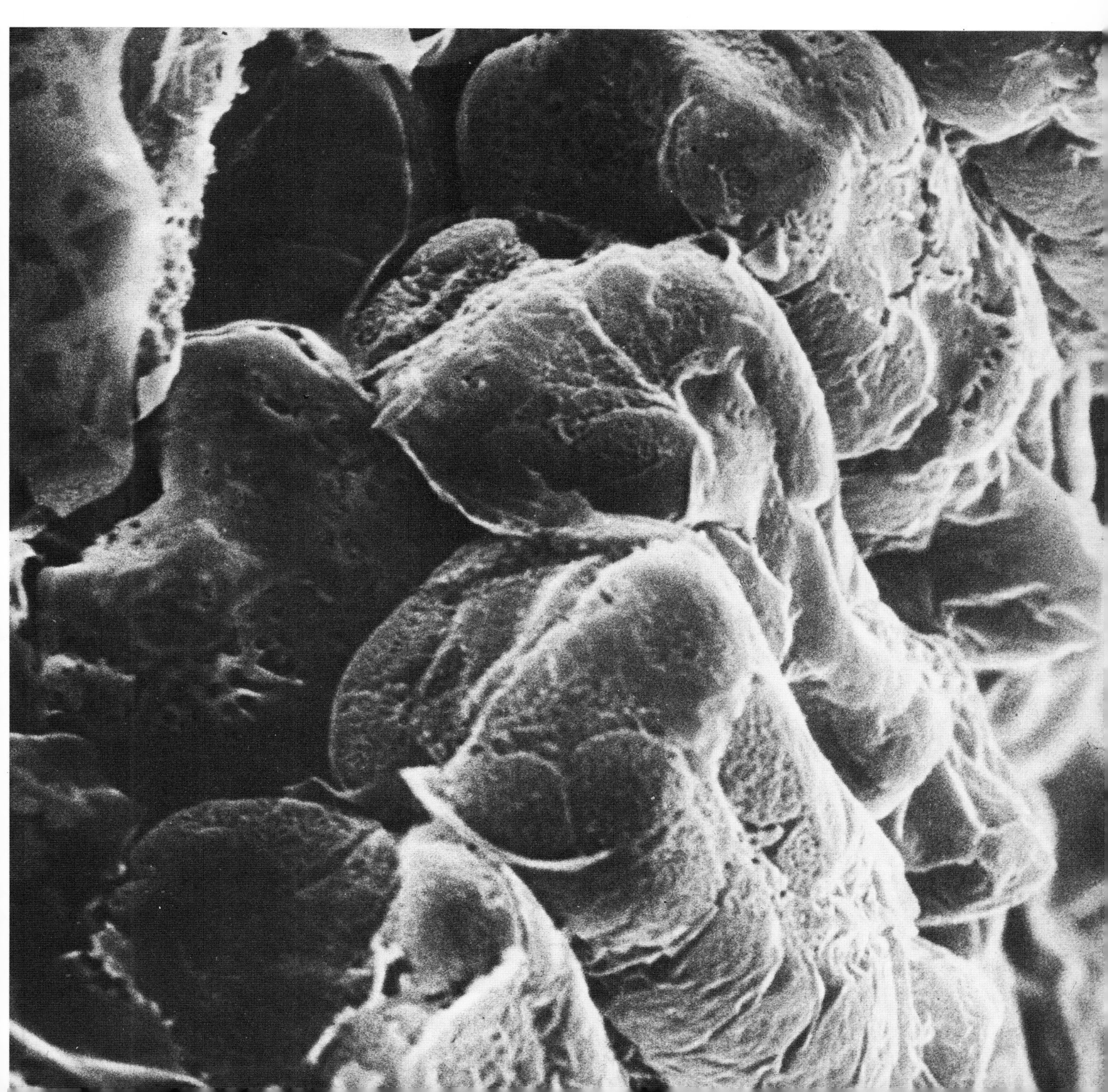

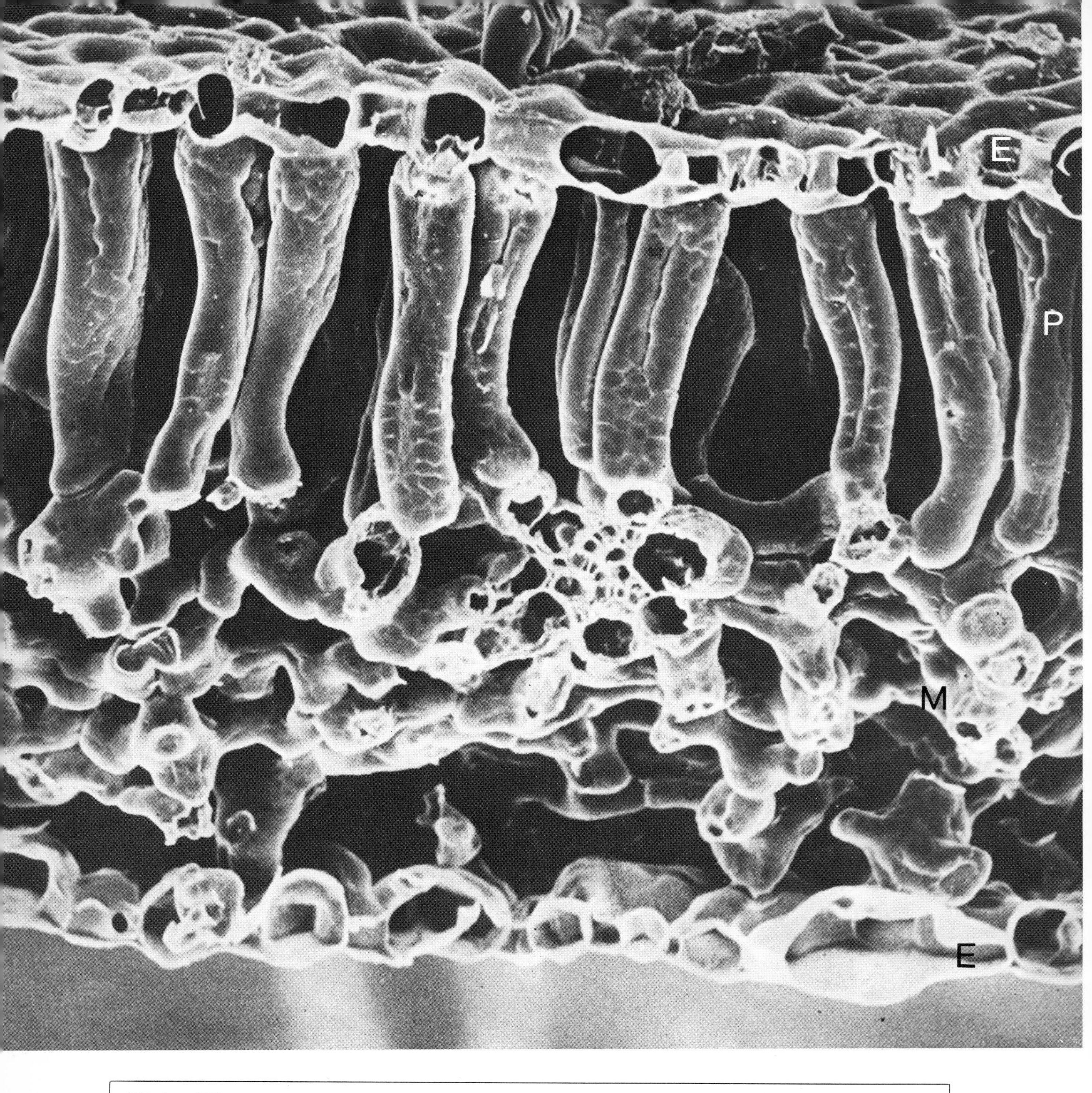

Plate 38 x420 | Inside the Leaf

Transverse view of a mature broad bean leaf. The internal organisation of cells in this leaf is characteristic of many plants, consisting of a layer of palisade cells (P) in the upper half of the leaf, spongy mesophyll cells (M) in the lower half and bounded on both sides by the epidermis (E). There are large air gaps between the palisade cells in the bean leaf, which is in contrast to the similar layer of cells in the tomato leaf. Most of the surface of the cells is exposed to the air and the area of cell wall exposed is about ten times the surface area of the leaf. The proportion of air volume to cell volume in a leaf can vary from 10% to 80% between different types of plants.

Plate 39 x1400 Inside the Leaf

Transverse view of a young broad bean leaf. The leaf was about two weeks younger than the leaf shown in Plate 38. The cells within the leaf are tightly packed together at this stage. The dramatic change in internal leaf structure with age has primarily been caused by an expansion in leaf area and by the extension of the palisade cells. The difference in final shape of the spongy mesophyll compared with the palisade cells indicates that the cell wall in the two types of cells may be quite different. The vascular system (V), containing the phloem and xylem can be seen in both leaves.

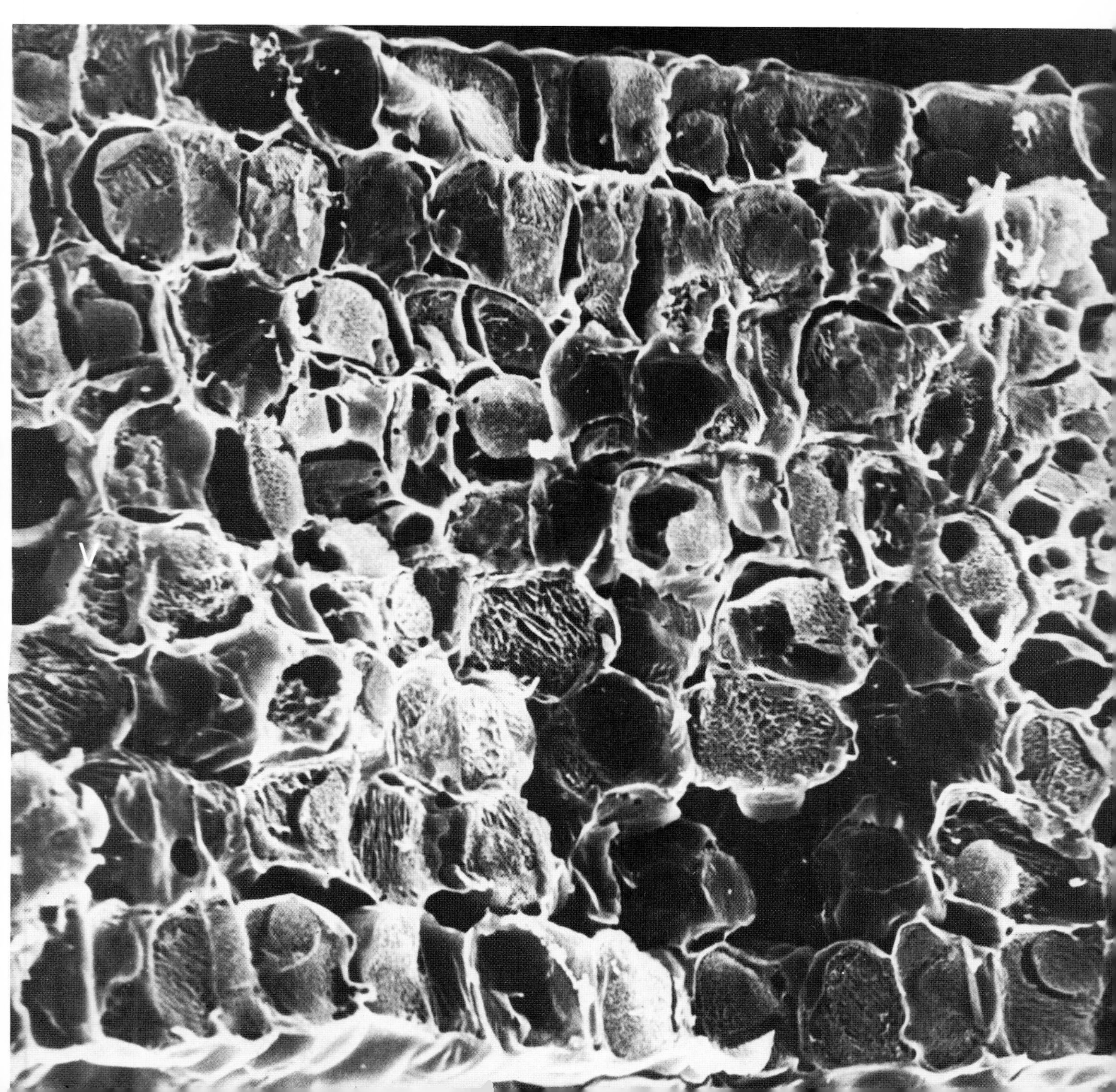

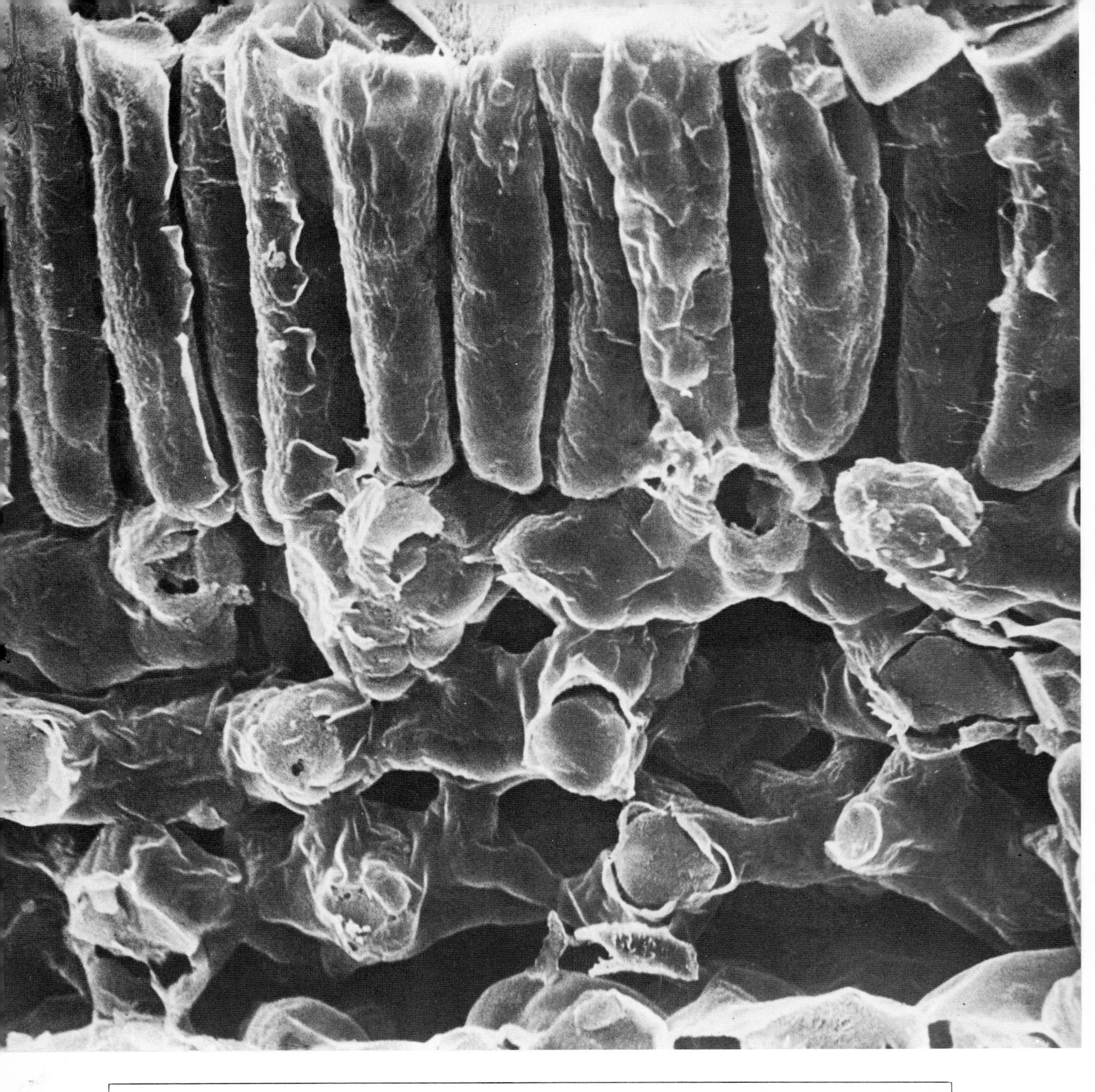

Plate 40 x1900 — Inside the Leaf

Transverse view of a leaf of cotton *(Gossypium hirsutum).* The arrangement of cells is similar to that of broad bean, with a palisade and spongy mesophyll layer; although the palisade cells are more tightly packed than in bean. The impression of chloroplasts on the walls can be seen in both types of cells, indicating that all mesophyll cells have chloroplasts. The epidermal cells however lack chloroplasts. The connections between cells can be seen in several places.

Plate 41 x490 — **Inside the Leaf**

Transverse view of a leaf of cucumber. This photograph illustrates the large hairs that arise from both the upper and lower epidermal cells. A stoma (S) with substomatal cavity can be seen in the upper epidermis. The transport of water into and through the leaf occurs in the xylem vessels and subsequently via the cell walls. The xylem vessels are located beside the phloem tissue and collectively these two types of transport systems are termed the vascular system (V).

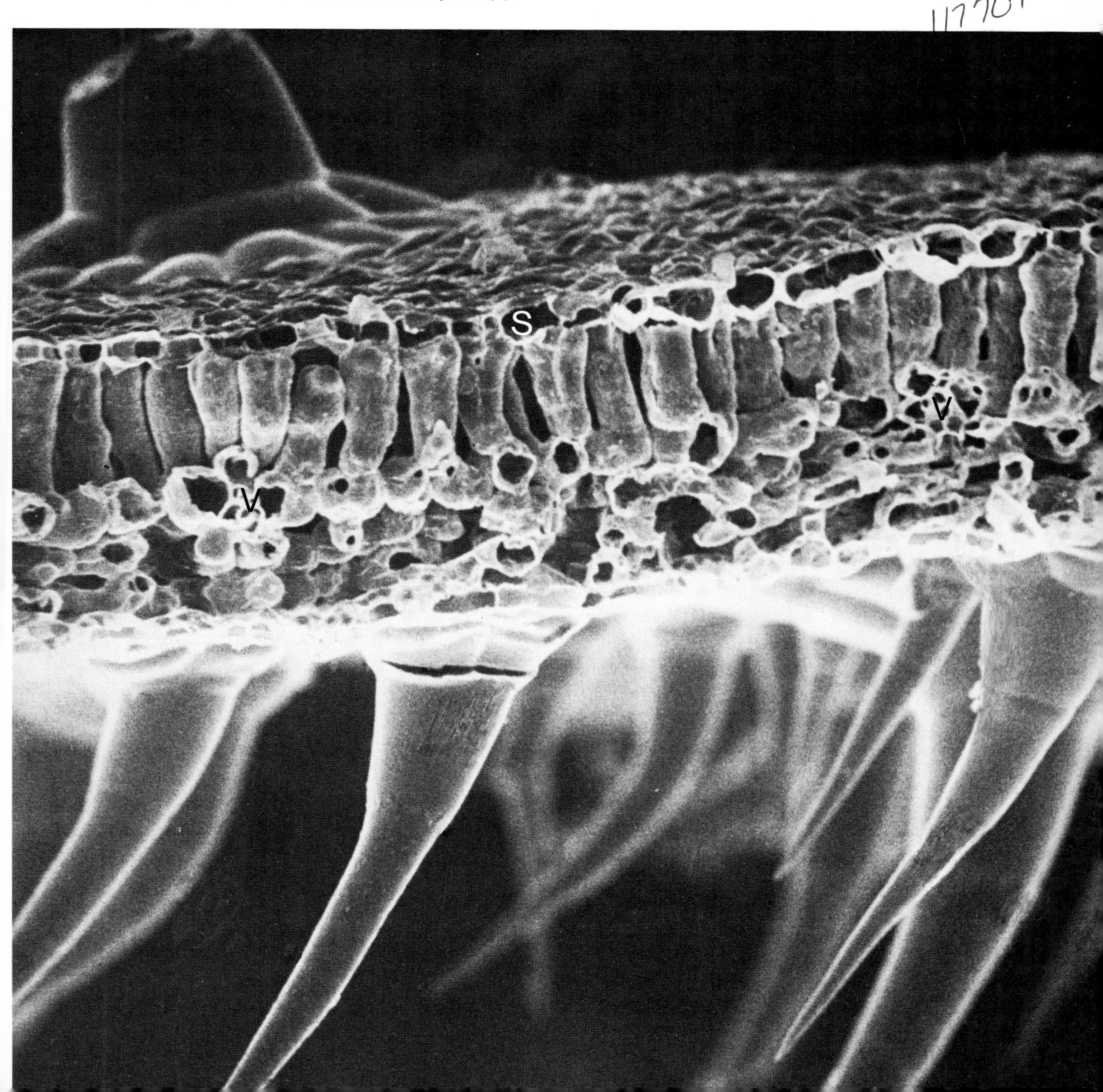

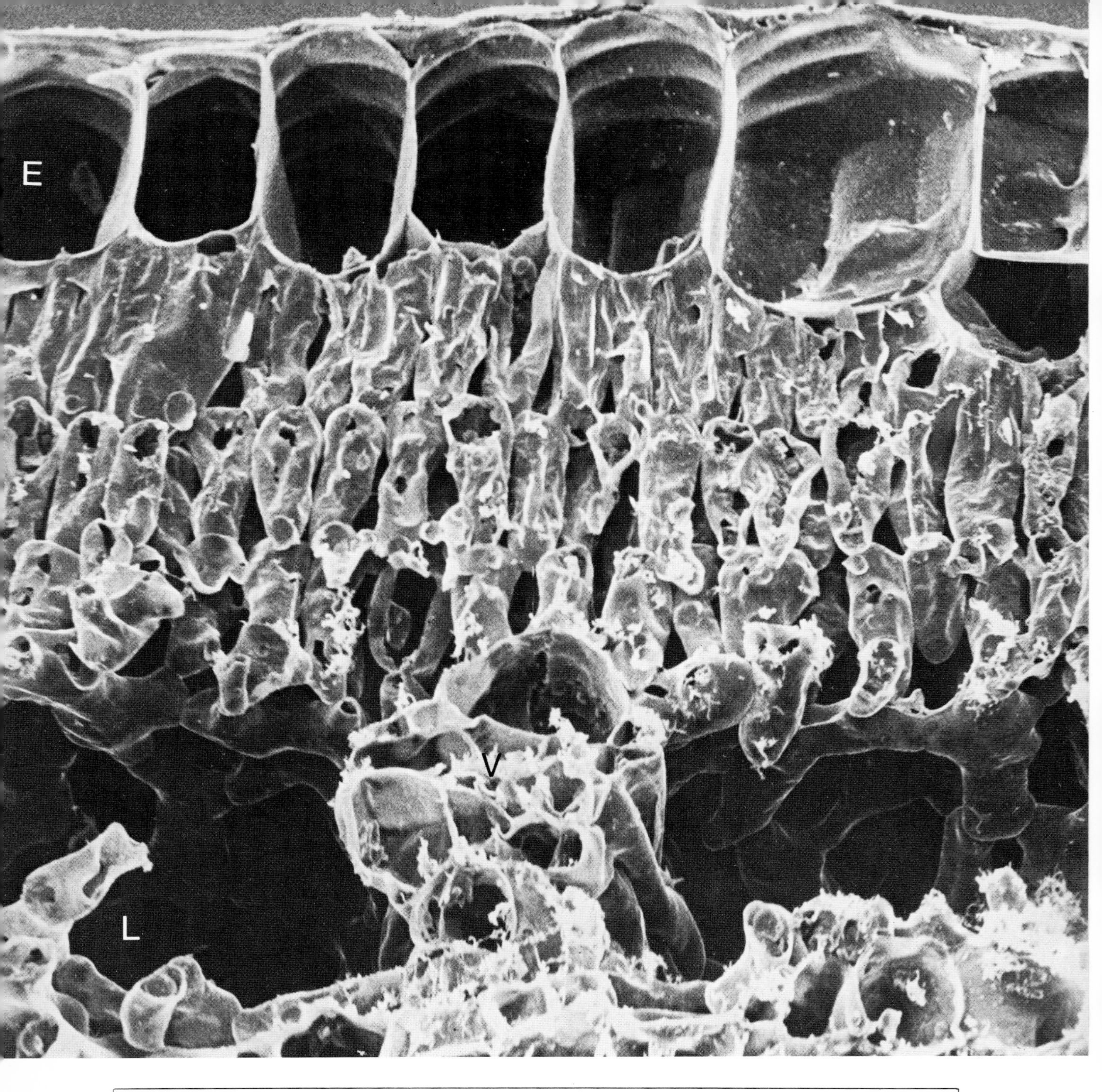

Plate 42 x1400 — Inside the Leaf

Transverse view of a leaf of banana. The internal anatomy of leaves is variable between species as well as within species. Banana has two or three layers of palisade cells with no distinct spongy mesophyll. In the region where the spongy mesophyll would normally be present there is a longitudinal air canal (L) which will allow rapid distribution of air within the leaf. The upper epidermal cells (E) are enlarged and have cell walls that give the appearance of being reinforced. The vascular (V) system is located between the air canals.

Plate 43 x370 — Inside the Leaf

Transverse view of a pine needle. The pine needle has several features which are adaptations for xerophytic environments. There is a layer of epidermal cells (E) with a heavy cuticle. The mesophyll (M) is not differentiated into palisade and spongy mesophyll cells and the innermost row of cells in the mesophyll have folds in the walls that protrude into the lumen. The endodermis (N) is a layer of thick walled cells that surround the vascular bundle (V) and transfusion tissue (T). There are two types of cells that make up the transfusion tissue, tracheids that have thick, lignified secondary walls and parenchyma cells. The physiological significance of the layer (N) is not known, but from its location within the leaf it is presumed to be involved in the transport of materials between the vascular tissue and the mesophyll.

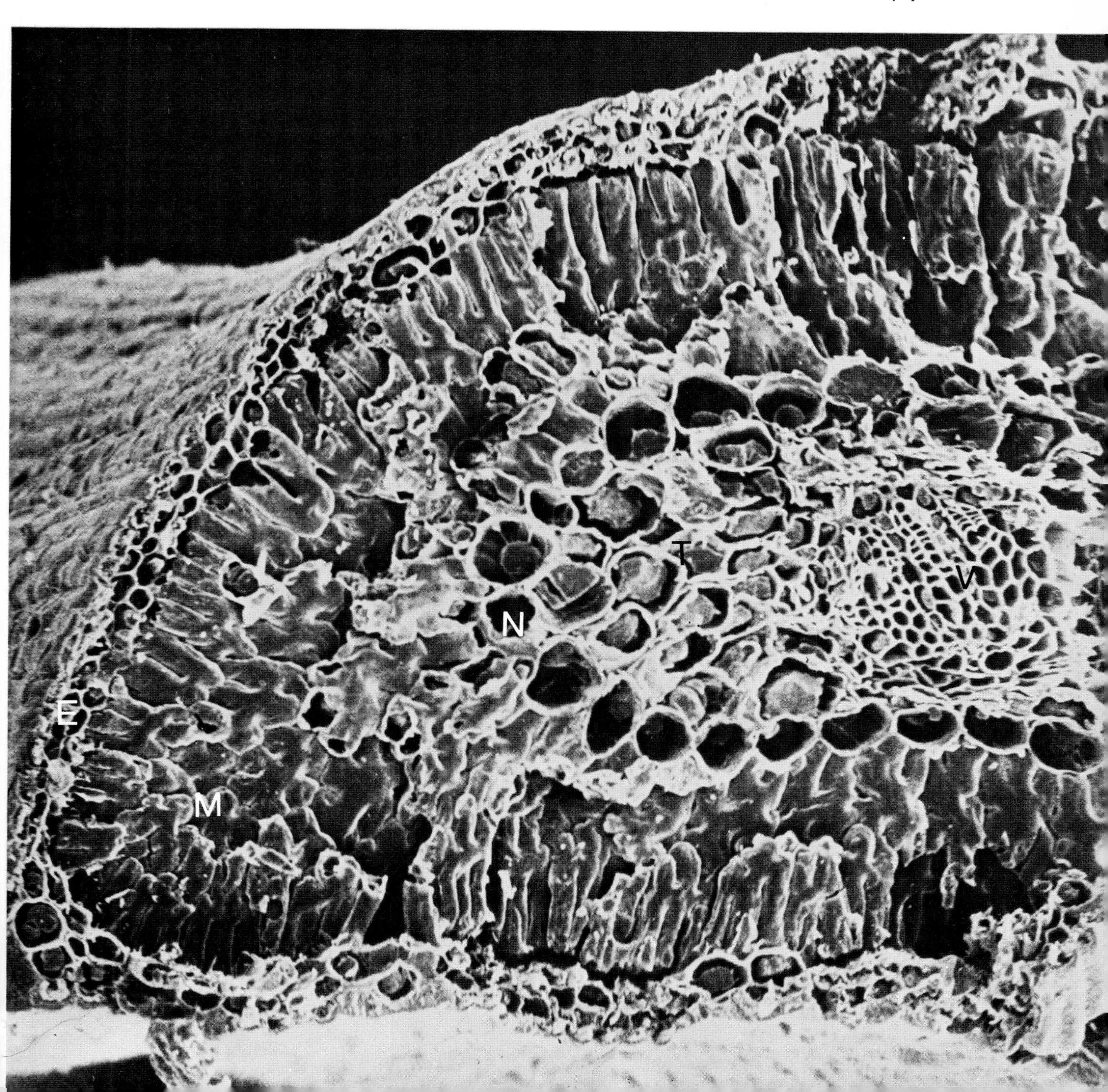

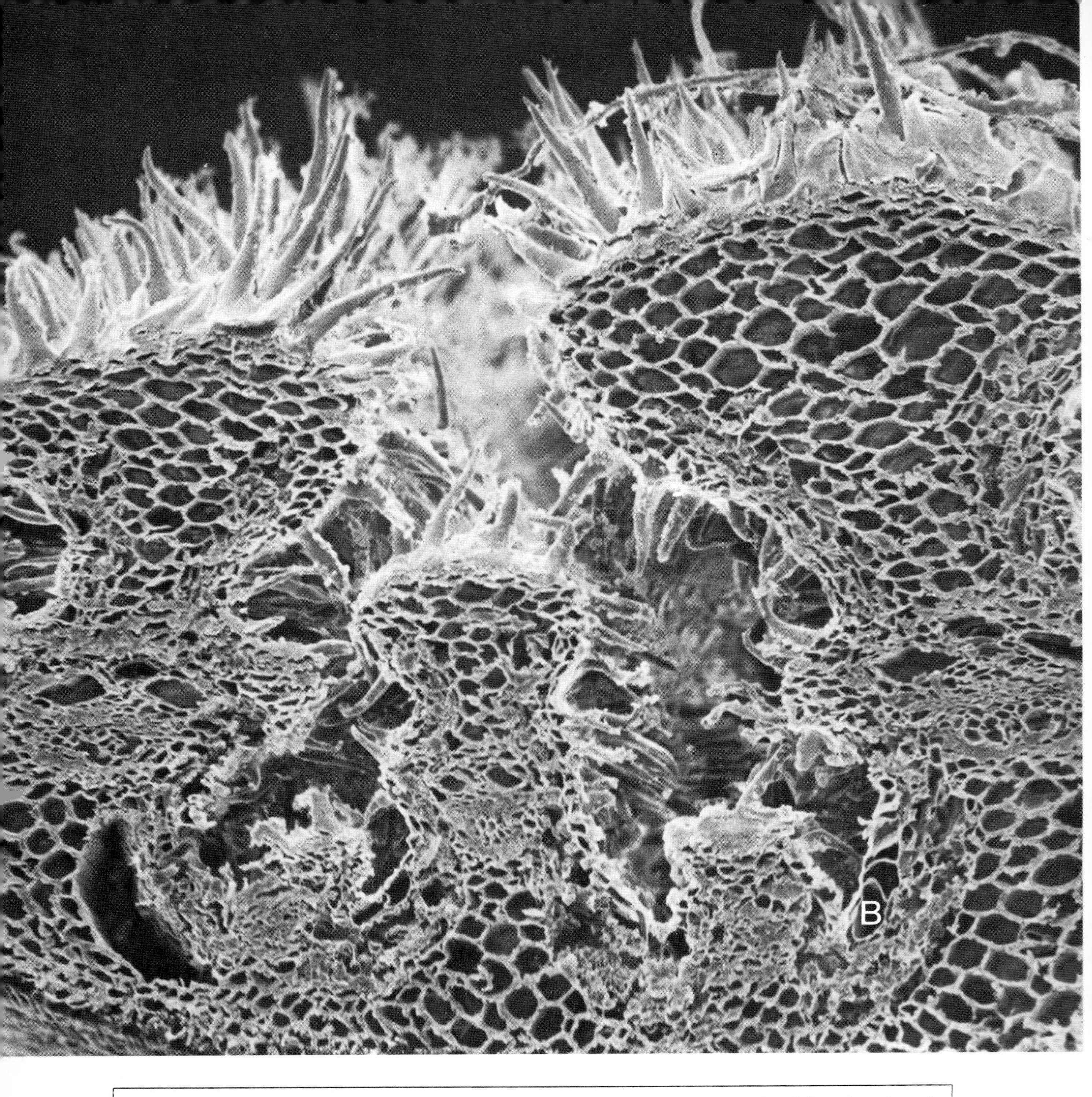

Plate 44 x270 — Inside the Leaf

Transverse view of a leaf of marram grass *(Ammophila arenaria).* The leaf of marram grass is covered with hairs which are an adaptation to a xerophytic habitat. A further adaption of this plant is to roll up the leaf during periods of water shortage. This reduces transpiration because of the reduction in the surface area to volume ratio and by increasing the path length for diffusion of water from the leaf to the air outside the leaf. Enlarged bulliform cells (B) are particularly involved in the folding of the leaf, although all cells participate in the action.

Plate 45 x350 Inside the Leaf

Transverse view of a pineapple leaf. The cells in this type of leaf are closely packed together and normally full of water. Most of the mesophyll cells have a rectangular shape but cells that traverse the air canal (C) have long narrow arms. The presence of the air canal increases the surface area of the cells that are exposed to intercellular air. Fibrous strands (F) are also present among the mesophyll cells.

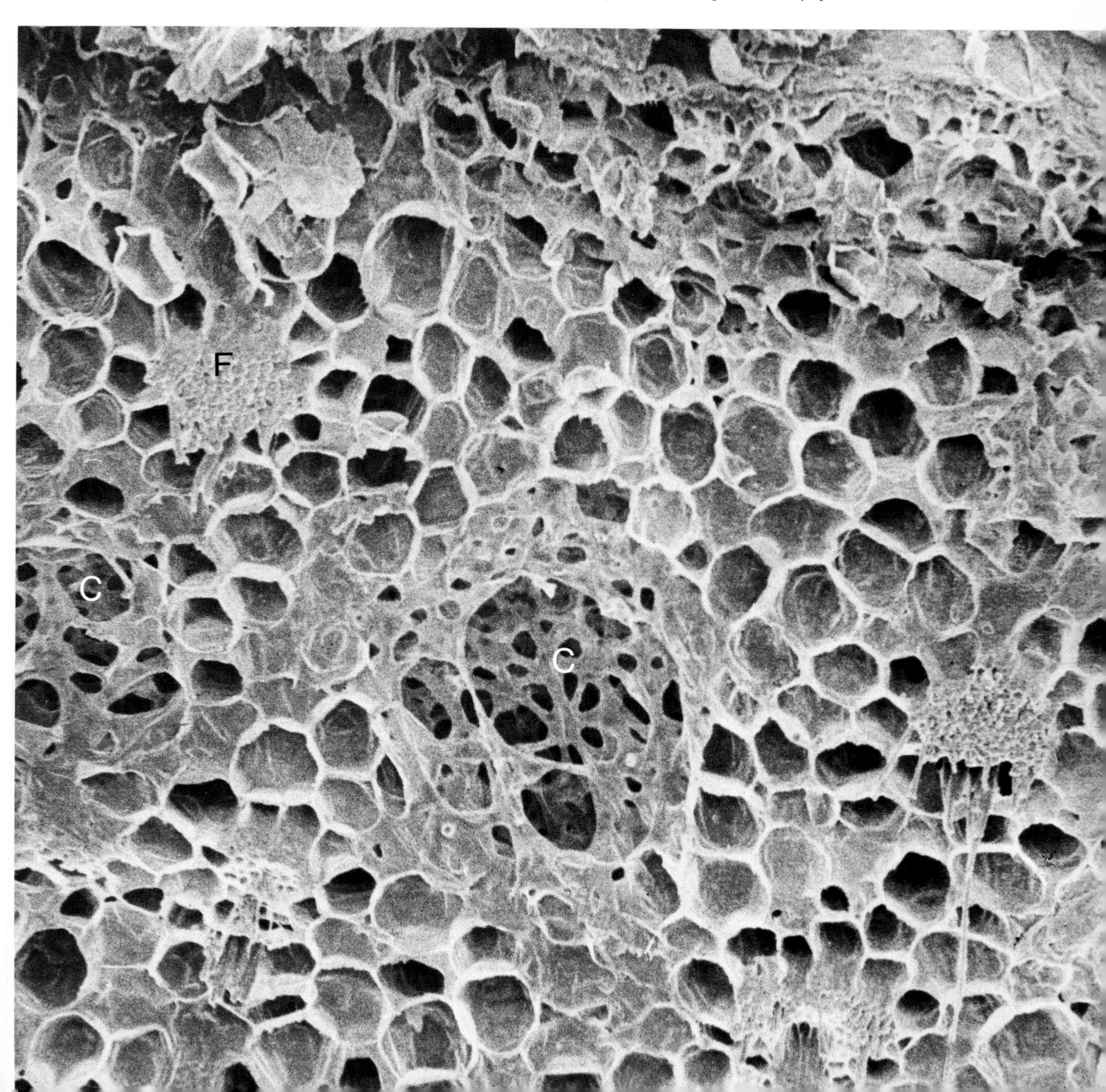

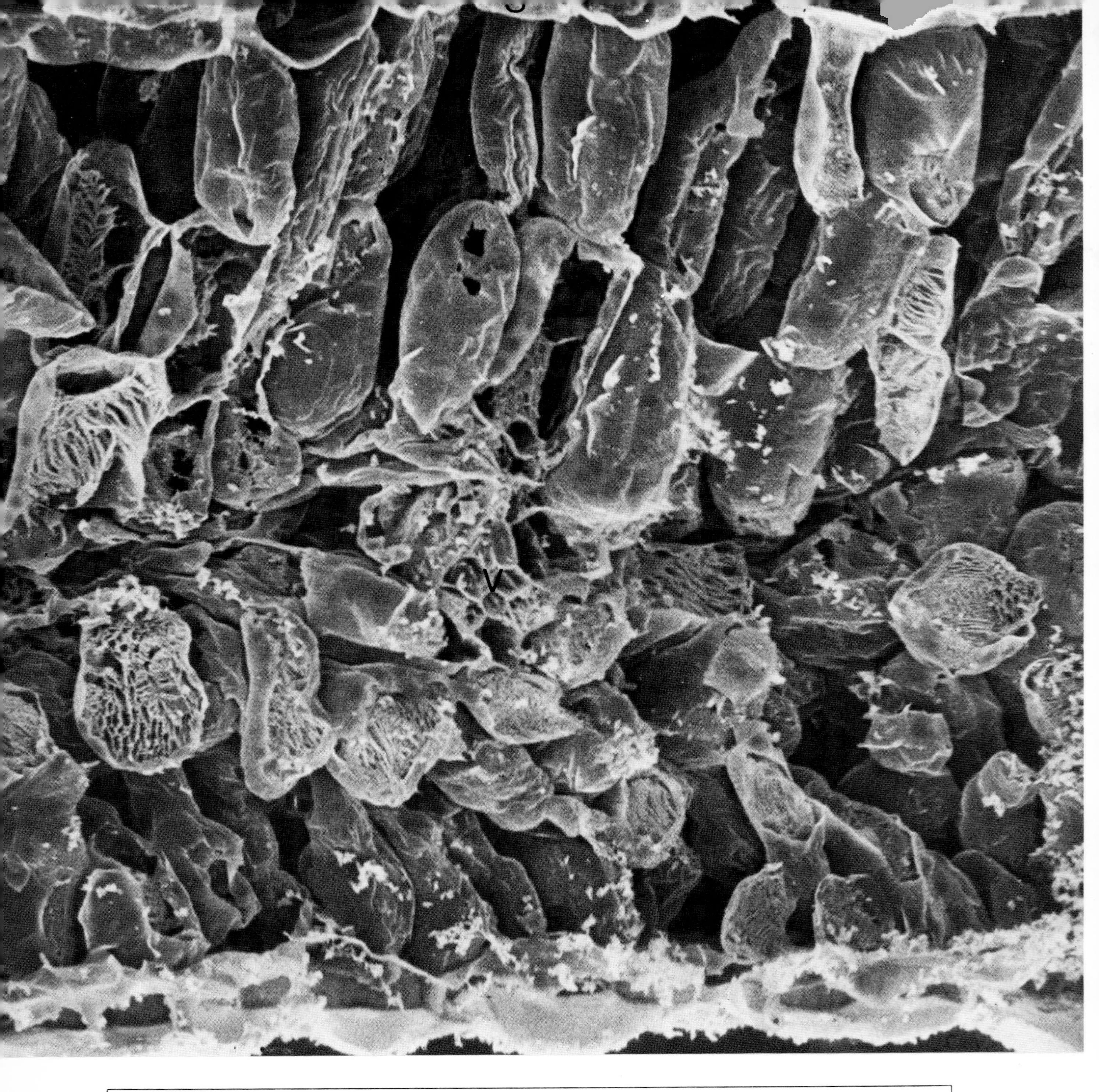

Plate 46 x420 | Inside the Leaf

Transverse view of an *Atriplex hastata* **leaf.** *Atriplex hastata* and *Atriplex spongiosa* are in the same genus, but each species has a different photosynthetic pathway and a different leaf anatomy. This later feature is shown in the two photographs shown here on opposing pages. In *Atriplex hastata* the palisade and spongy mesophyll layers are present, although not as well developed as in cotton or bean. Note also the stoma (S) and vascular tissue (V). This plant has the Calvin or C_3 type photosynthetic pathway which involves the fixation of the CO_2 in the chloroplasts by the enzyme ribulose-1:5-diphosphate carboxylase and the production of phosphoglyceric acid. At the same time as photosynthesis is occurring, CO_2 is being produced within the cells (photo-respiration) and although some of the CO_2 is refixed, much of it escapes from the leaf.

Plate 47 x630 | **Inside the Leaf**

Transverse view of a saltbush leaf. The dominant features of the internal organisation of the cells in a saltbush leaf are the presence of the bundle sheath parenchyma cells (B) surrounding the xylem and phloem and the mesophyll cells (M) radially arranged round the bundle sheath and vascular tissue (V). Anatomically this is distinct from *Atriplex hastata* and the biochemical differences between the species are in the initial reactions of photosynthesis. In saltbush the enzyme involved in the initial fixation of CO_2 is phosphoenolpyruvate carboxylase (PEP carboxylase) and the immediate products of this reaction are malate and aspartate. Species with this characteristic are termed C_4 type plants. The exact location of PEP carboxylase within the cell is not known, but it is thought to be on the outside of the chloroplast or free in the cytoplasm of the mesophyll cell.

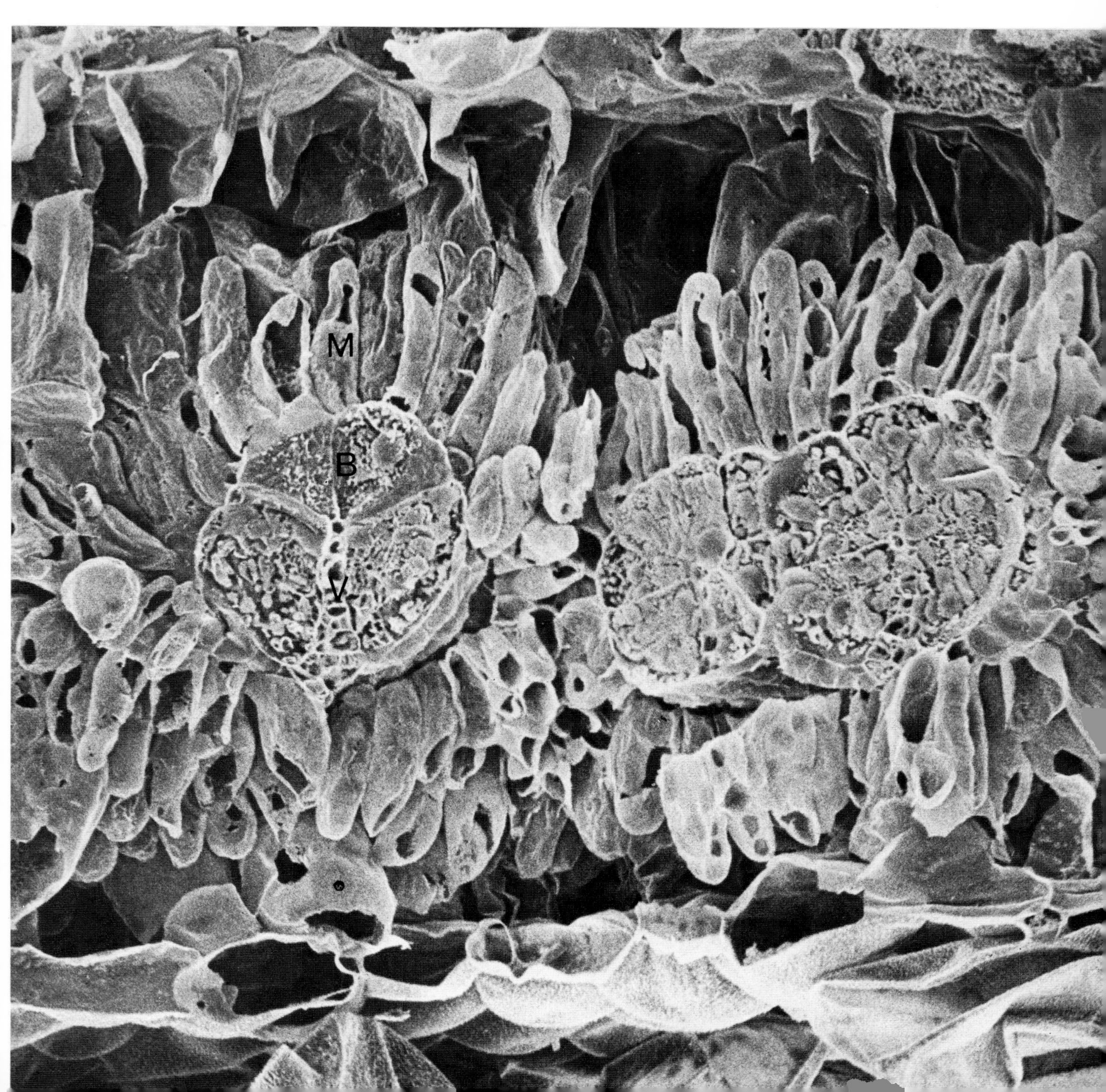

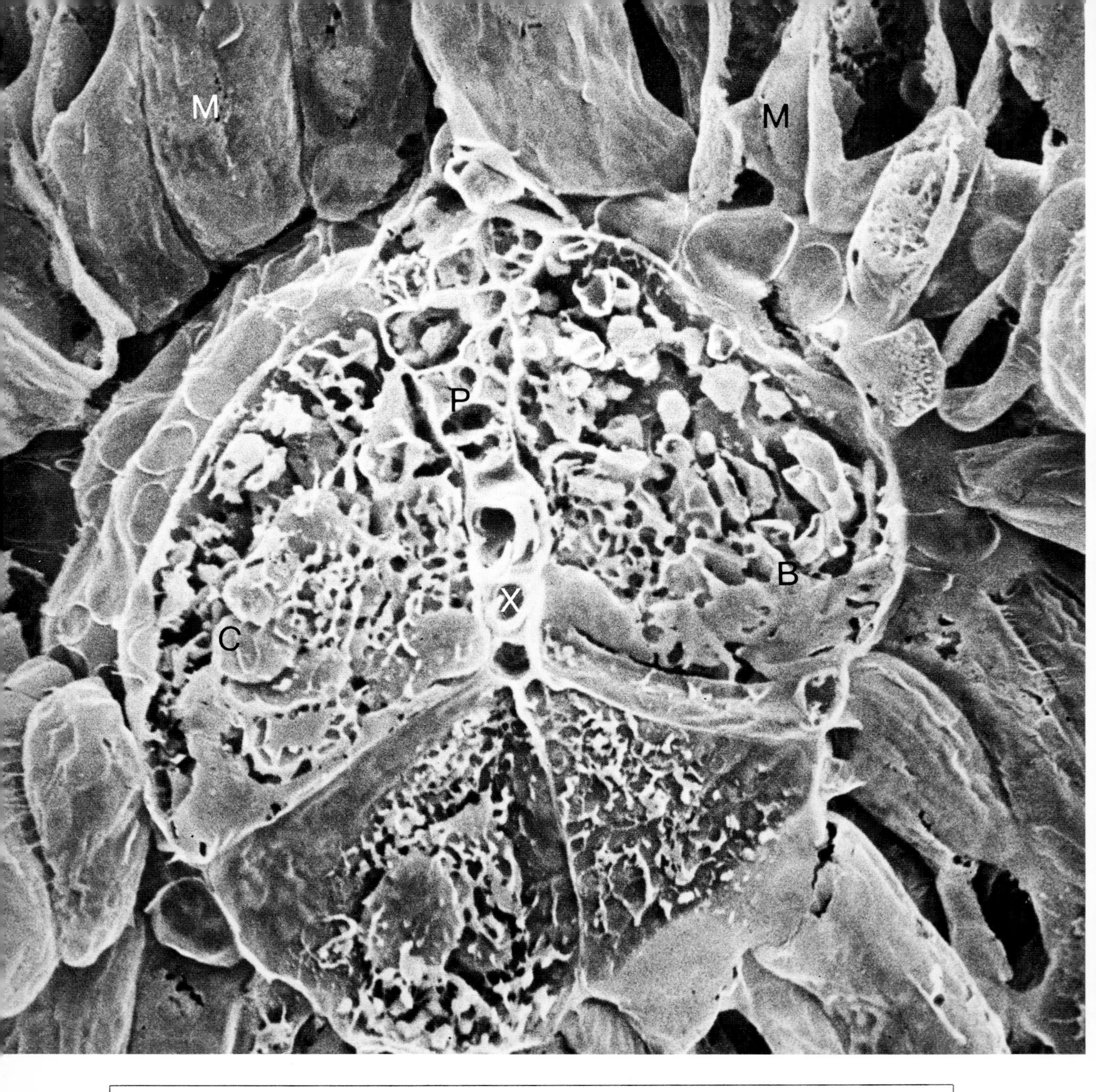

Plate 48 x2300 Inside the Leaf

Transverse view of bundle sheath parenchyma cells in a leaf of saltbush. The anatomy of the mesophyll and bundle sheath cells in C_4 type plants are quite distinct and these cells also have specialised biochemical functions. The carbon dioxide is fixed in the mesophyll cells (M) by PEP carboxylase and transported as malate through plasmodesmata into the bundle sheath cells (B). In this latter cell, CO_2 may be released and refixed by RuDP carboxylase and subsequently transformed into more complex carbon compounds by the C_3 type reactions. Some of the compounds remain in the cell and can accumulate, for example starch, while other compounds are transported, as sucrose, throughout the plant in the phloem (P). The xylem (X) and phloem divide two bundle sheath cells. The chloroplasts (C) are scattered through the bundle sheath cells.

Plate 49 x790 — Inside the Leaf

Transverse view of a *Gomphrena globosa* **leaf.** *Gomphrena globosa* is also a C_4 type plant but has a slightly different anatomy from saltbush. Different features include the less well developed mesophyll cells on the lower half of the leaf, the bundle sheath cells that completely surround the xylem and phloem and the chloroplasts in the bundle sheath are aggregated towards the vascular tissue and not spread throughout the cell as in saltbush. Many other species have the C_4 type characteristics in anatomy and biochemistry, including the economically important plants maize, sugar cane, sorghum, paspalum and several other tropical grasses. These species generally have higher rates of photosynthesis in high light than C_3 type plants, which is most likely due to the higher affinity for CO_2 of PEP carboxylase compared with RuDP carboxylase. However other features that are involved include the radial arrangement of the mesophyll cells and the large cell surface area available to the intercellular air. Furthermore if PEP carboxylase is free in the cytoplasm, or on the outer membrane of the chloroplasts, the path length for CO_2 diffusion into the cell will be shorter.

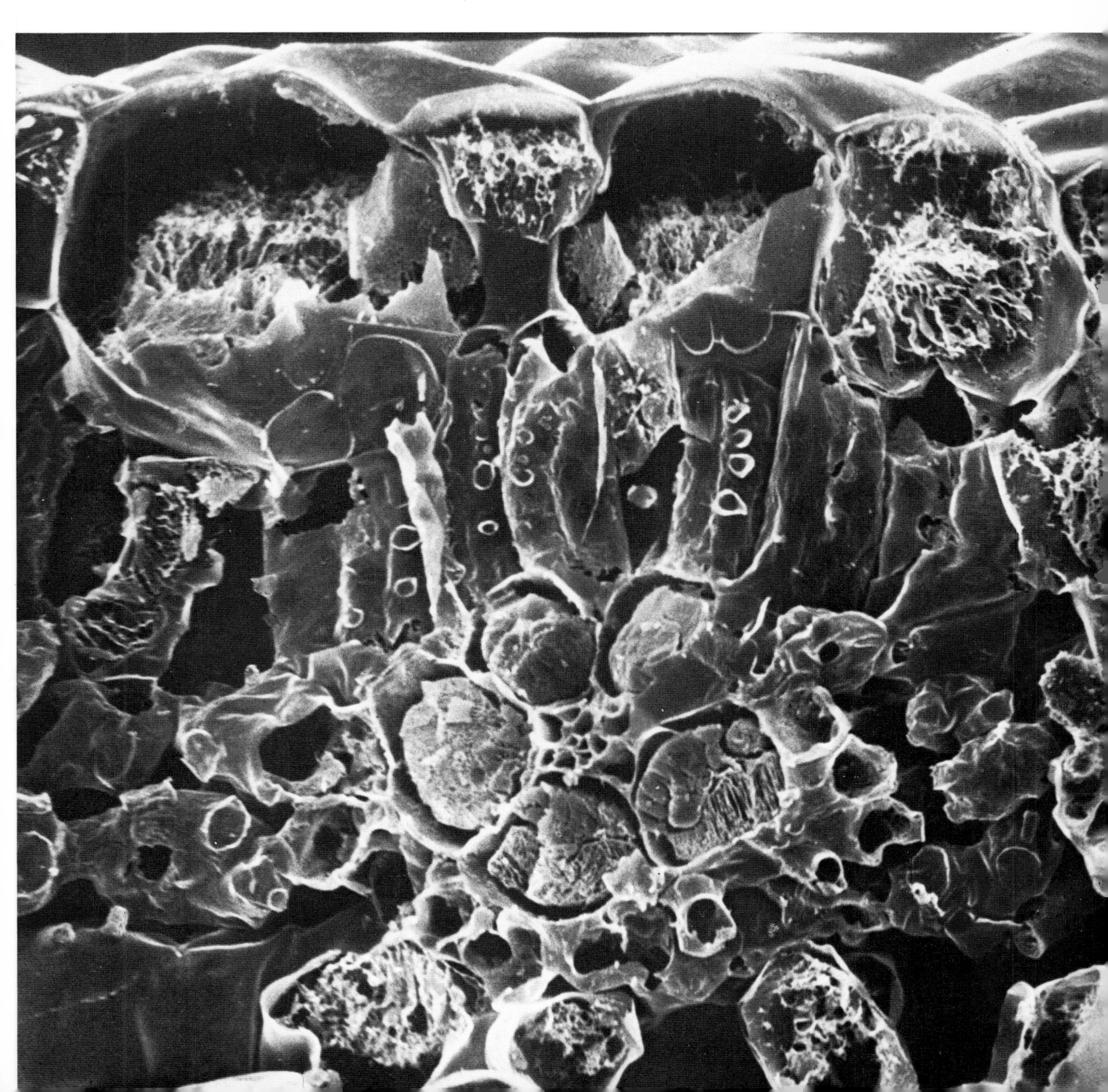

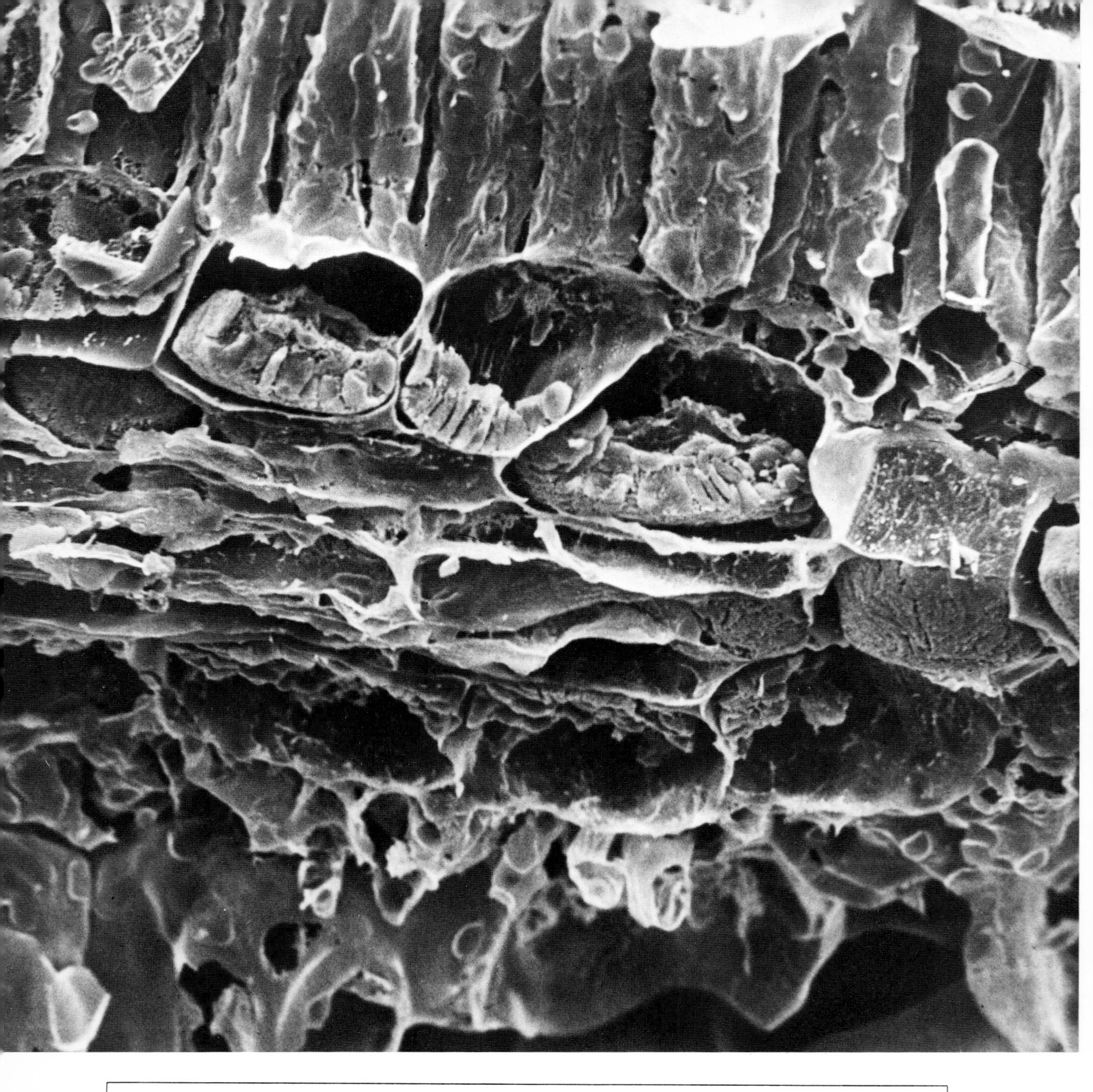

Plate 50 x1200 Inside the Leaf

Longitudinal view of a *Gomphrena globosa* **leaf.** Several mesophyll cells feed into a single bundle sheath cell and it can be seen that the chloroplasts in the bundle sheath are organised along the phloem. Under normal conditions no CO_2 can be detected as being evolved in C_4 type plants, which suggests photorespiration may not be present. There is no evidence for the glycolate pathway of photorespiration in the mesophyll cells but some compounds associated with this pathway have been detected in the bundle sheath cells. As CO_2 production cannot be detected in these plants, it has been suggested that either CO_2 is not produced or if produced it is recycled within the cells and not released to the air. It has also been suggested that the cell wall surrounding the bundle sheath cells is impermeable to gases and hence traps any CO_2 produced, either by photorespiration or the C_4 pathway, which is then refixed by RuDP carboxylase. Alternatively the oxygen concentration in these cells may be low, which would also reduce photorespiration.

Plate 51 x790 — Inside the Leaf

Transverse bundle in a *Gomphrena globosa* leaf. Transverse bundles connect the major vascular bundles that ramify throughout the leaf and it can be seen that these bundle sheath cells also contain chloroplasts. Xylem elements (X) are also present and they supply the adjacent cells with water. Normally the transport of water from the xylem elements occurs via the cell walls, but it has been suggested that the bundle sheath cell wall is impermeable to water. If this is so then the water supplying the mesophyll and epidermal cells must pass through the bundle sheath cell and plasmodesmata.

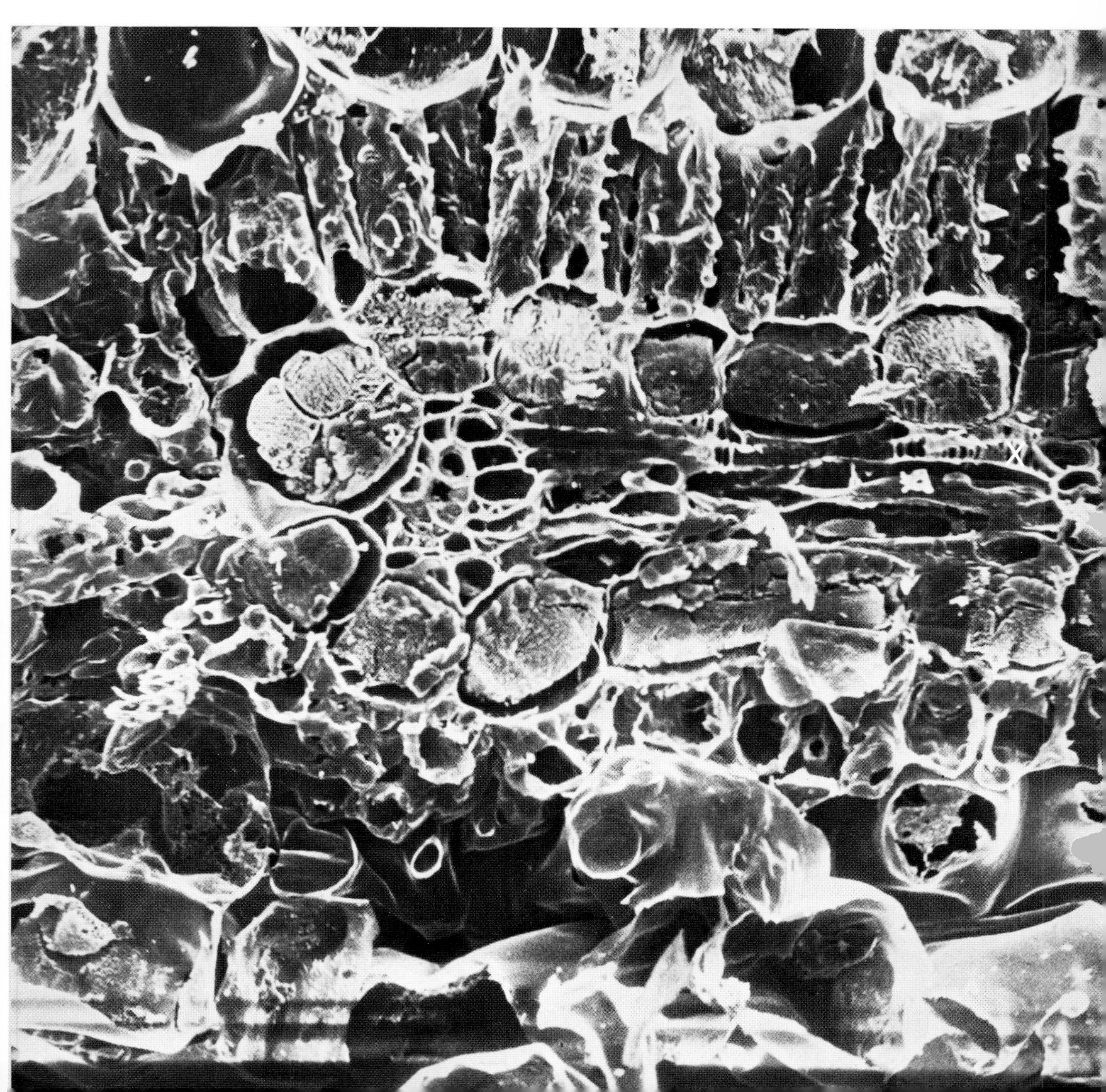

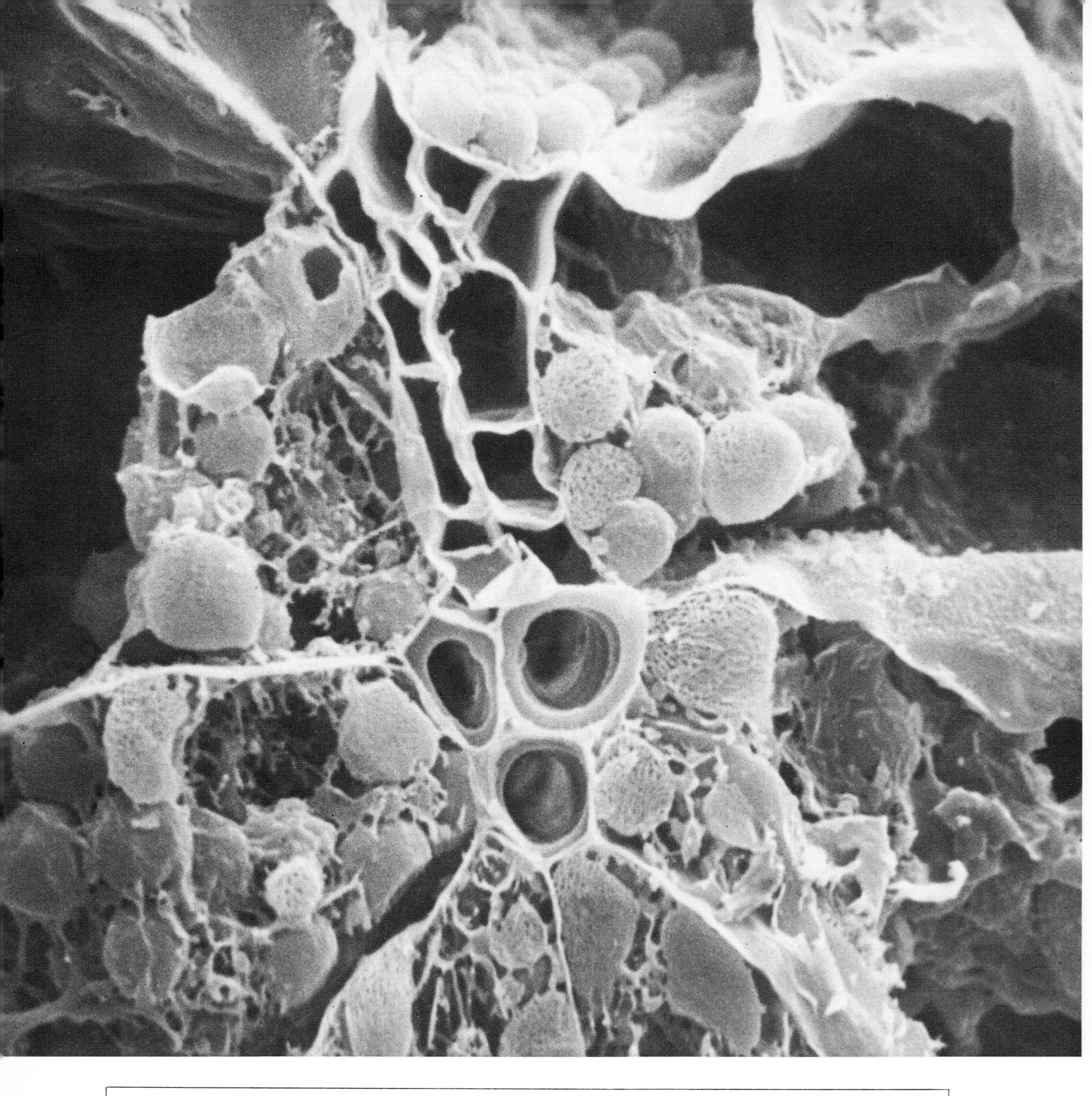

Plate 52 x3500 — Inside the Leaf

Chloroplasts in the bundle sheath cells of portulaca *(Portulaca sp.).* In this example the chloroplasts are clustered towards the inner side of the bundle sheath. A major disadvantage of the freeze drying technique for preparing tissue to be used in the S.E.M. is that water is removed from the cells. This gives the impression that most of the volume of a cell is empty, whereas under normal conditions the cells are filled with fluid. The cell normally consists of the plasmalemma between the chloroplasts and the cell wall, a thin strip of cytoplasm containing the chloroplasts, the tonoplast and the vacuole which is primarily water. There are several other organelles which are present in the cytoplasm along with the chloroplasts and these are the nucleus, mitochondria and peroxisomes. The last named organelle is associated with some of the reactions of photorespiration while the mitochondria contain enzymes associated with the Krebs cycle of respiration and a respiratory chain of compounds which utilise nicotinamide adenine dinucleotide phosphate (NADP) in the reduced form $NADPH_2$ to produce adenosine triphosphate (ATP).

Plate 53 x6700	Chloroplasts

Chloroplasts aligned along the phloem in portulaca. A feature of the C_4 type plants is that the bundle sheath cells that are associated with the production of sucrose are closely associated with the phloem cells. In C_3 type plants the average distance between a chloroplast and the phloem is about two and a half cells, whereas in C_4 type plants it is about one cell. On the basis of the geometry of the leaf, C_4 type plants have an advantage over C_3 type plants in the rapidity with which carbon can be moved from the photosynthesising cells into the translocation system.

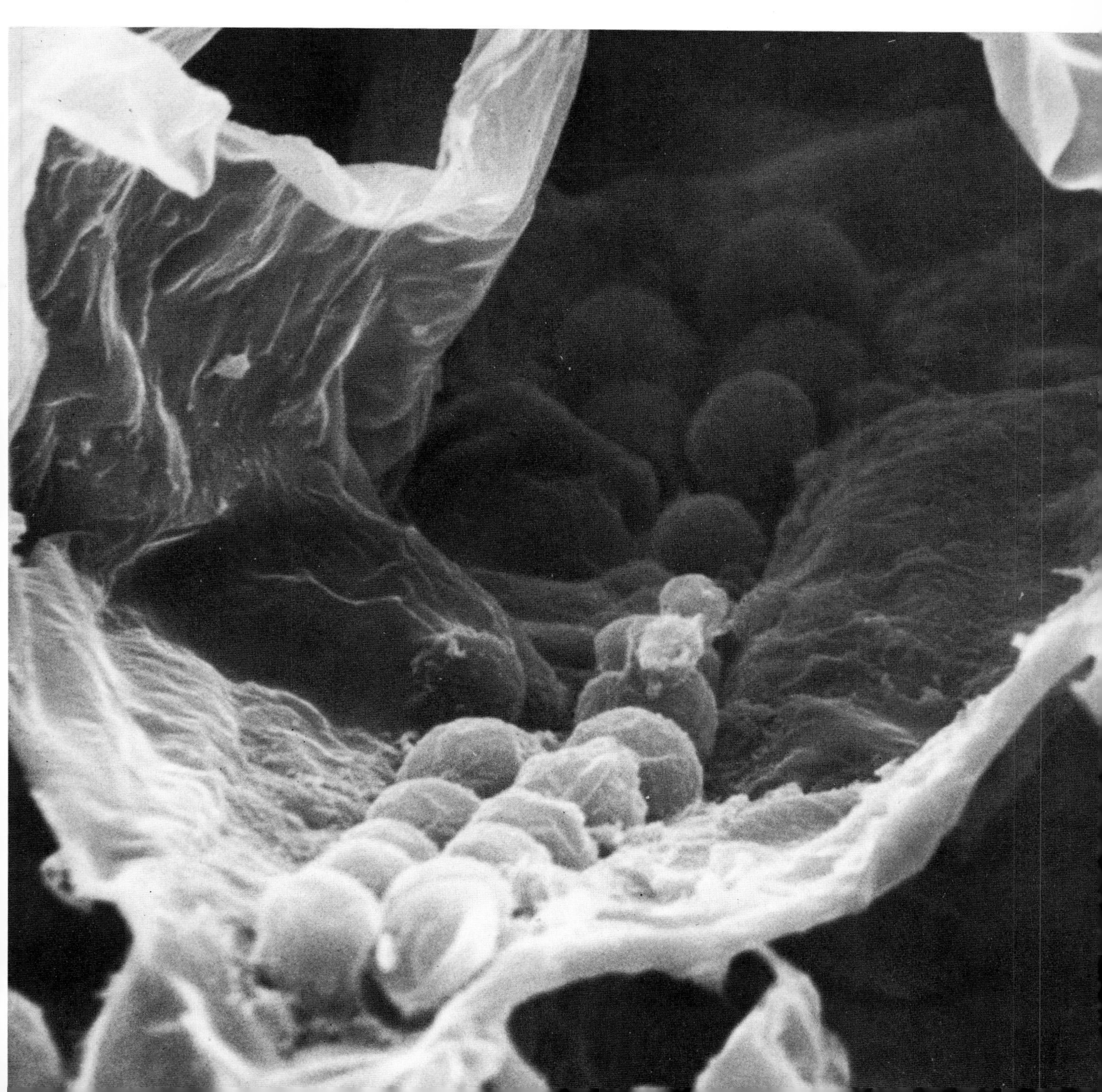

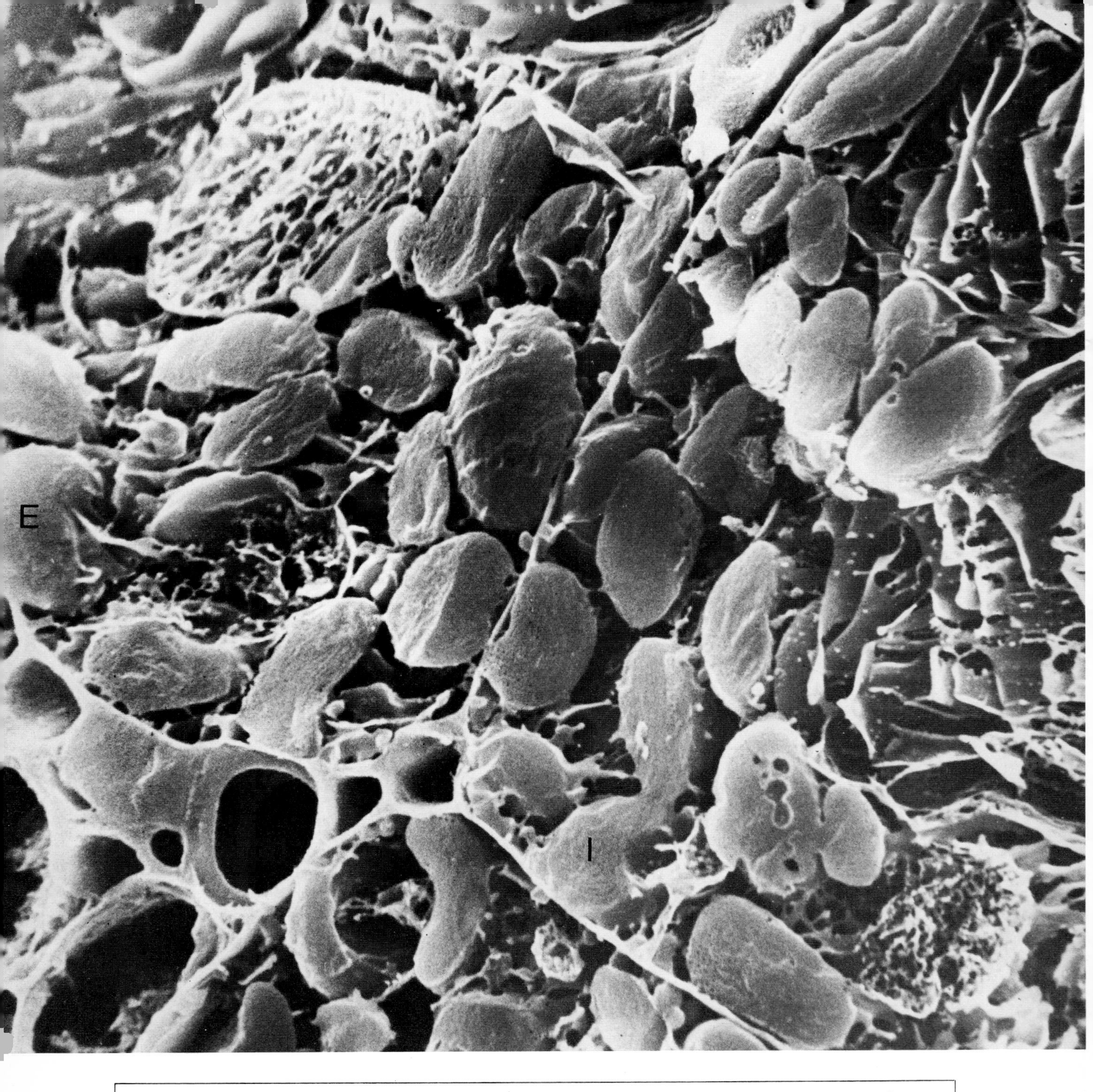

Plate 54 x2300 — Chloroplasts

Chloroplasts in the bundle sheath cells of portulaca. The chloroplast is surrounded by a membrane, the outer envelope, which controls the passage of ions and chemical substances between the cytoplasm and the inside of the chloroplast. Thus the chemical environment inside the chloroplast is different from the cytoplasm which in turn is different from the vacuole. The surface of the outer envelope (E) can be seen in this photograph but some chloroplasts have broken in half so that the view that can be seen is of the inside of the chloroplast (I). The inside of these chloroplasts are dense, illustrating that they are filled with substances other than water and it is known from using other techniques that there are numerous membranes within the chloroplasts. To view these membranes it is necessary to have better resolution than can be obtained with the S.E.M. at present, although faint lines running through the chloroplast (I) would be the internal or thylakoid membrane.

Plate 55 x3600 Chloroplasts

Chloroplasts in the bundle sheath cells of *Gomphrena globosa.* The green colour of plants is due to the pigment chlorophyll that is located within the chloroplasts. Chloropyhll is located in the internal membranes of the chloroplast and is responsible for the absorption of the light energy which is converted to chemical energy in the form of high energy phosphate bonds. The thylakoid membranes are the site for the light reactions and the end result of these reactions is the production of ATP and $NADPH_2$. Subsequently these chemicals are used within the cell in many ways but in photosynthesis they are used to reduce CO_2 to phosphoglyceraldehyde.

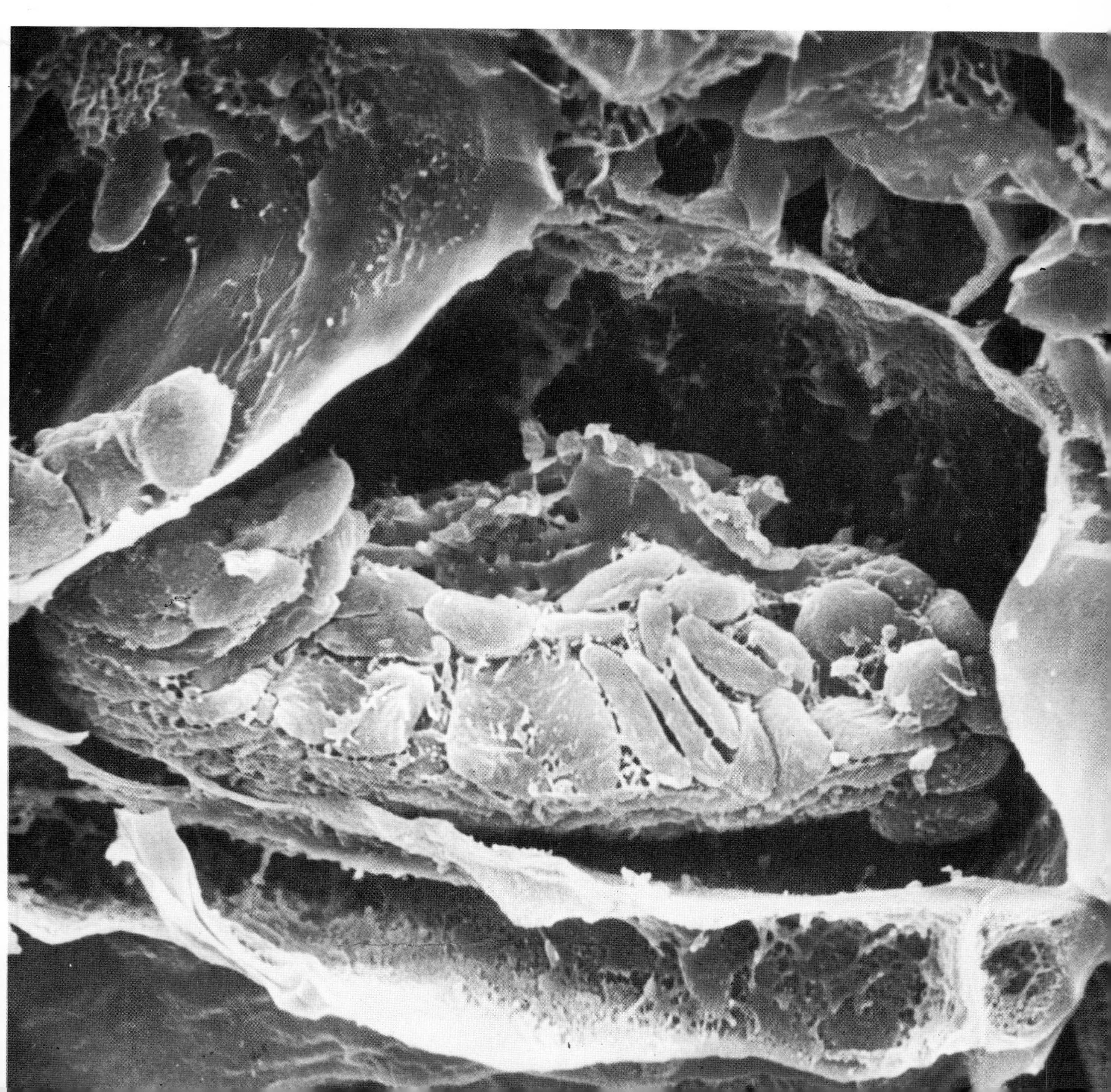

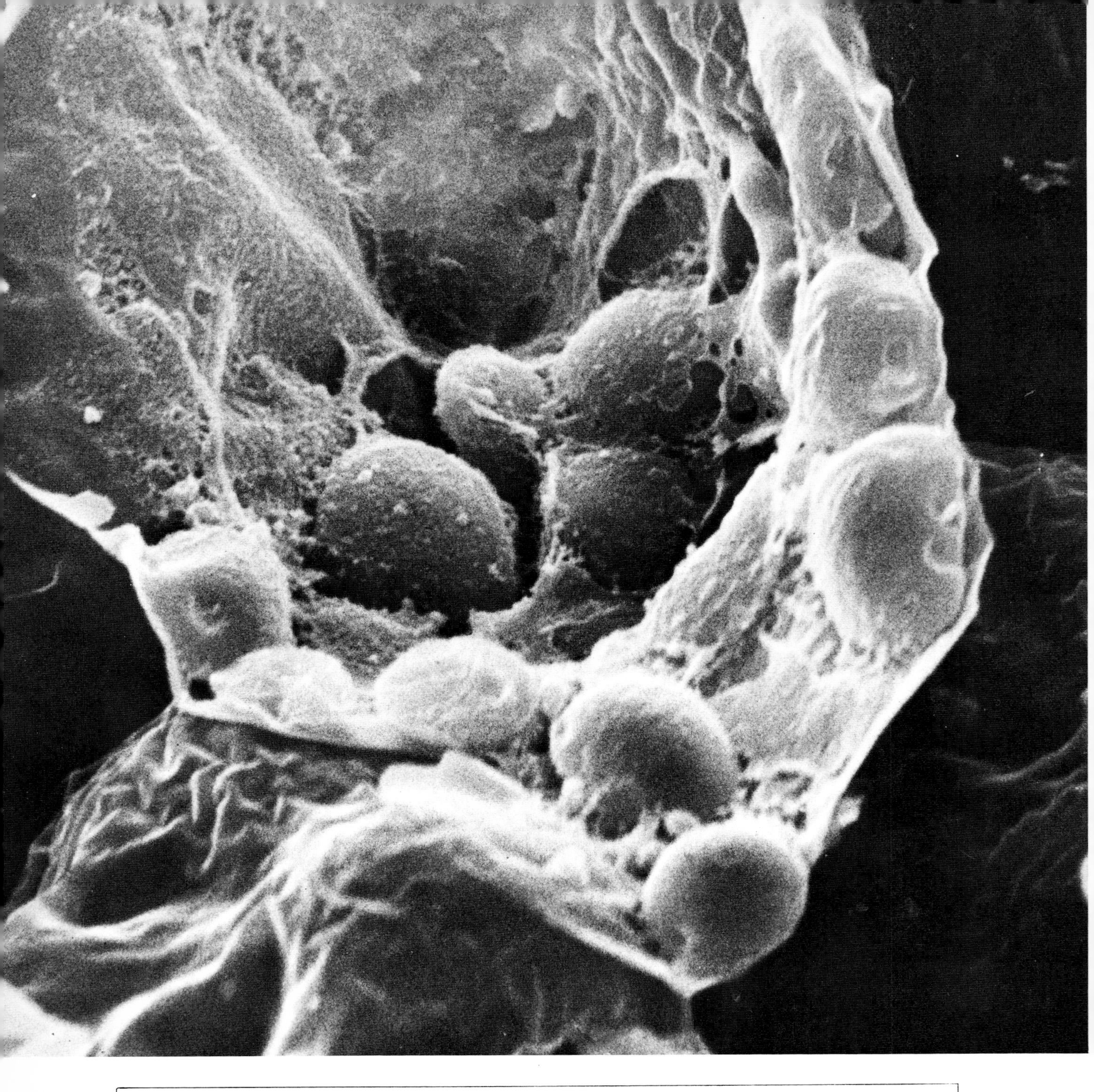

Plate 56 x5550 Chloroplasts

Chloroplasts in a mesophyll cell of cucumber. The chloroplasts in mesophyll cells such as this one are closely associated with the cell wall over most of the cell. Views as in this photograph indicate the chloroplasts are static but under some conditions chloroplasts can move round inside the cell. The chloroplast is the end point in the transfer of CO_2 along the tortuous pathway from the outside air to being the substrate for photosynthesis. Under many conditions this pathway restricts the supply of CO_2 to the chloroplast so that the rate of photosynthesis is regulated by the amount of CO_2 that is available in the cell. Consequently by increasing the amount of CO_2 in the air round plants artificially it is possible to get more CO_2 into the chloroplast and photosynthesis and the growth of the plant is increased. However there are many other processes involved in growth apart from photosynthesis so that an increase in photosynthesis may not automatically mean an increase in growth rate.

Plate 57 x50 Xylem

The root tip of a radish *(Raphanus sativus)* **seedling.** The leaves require an adequate supply of water to allow plant growth to proceed normally. A lack of water in the leaves directly interferes with enzymes, cell membranes and biochemical reactions and indirectly will prevent cell expansion and close stomata because of a loss of turgor. Water is drawn into the plant through the root which also functions in the absorption of nutrients from the soil and as an anchor for the plant. In the photograph of a root of a radish seedling there are two distinct regions, the root tip region free of hairs, and the root hair region. The root hairs increase the absorptive area of the root but most root surfaces are permeable to water. The root grows by cell division, cell growth and cell expansion and as the cells mature they differentiate into different types of cells with specific characteristics.

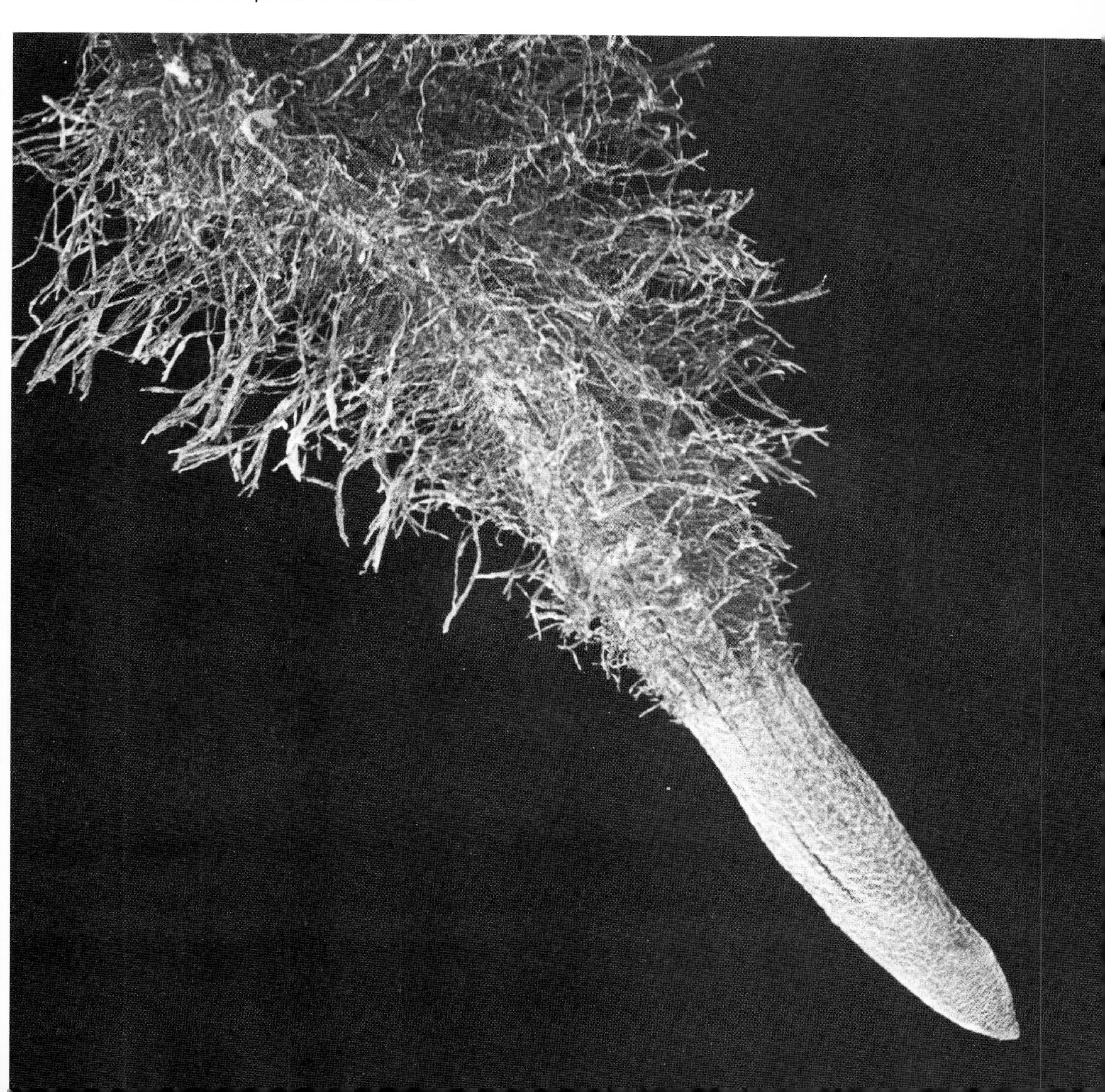

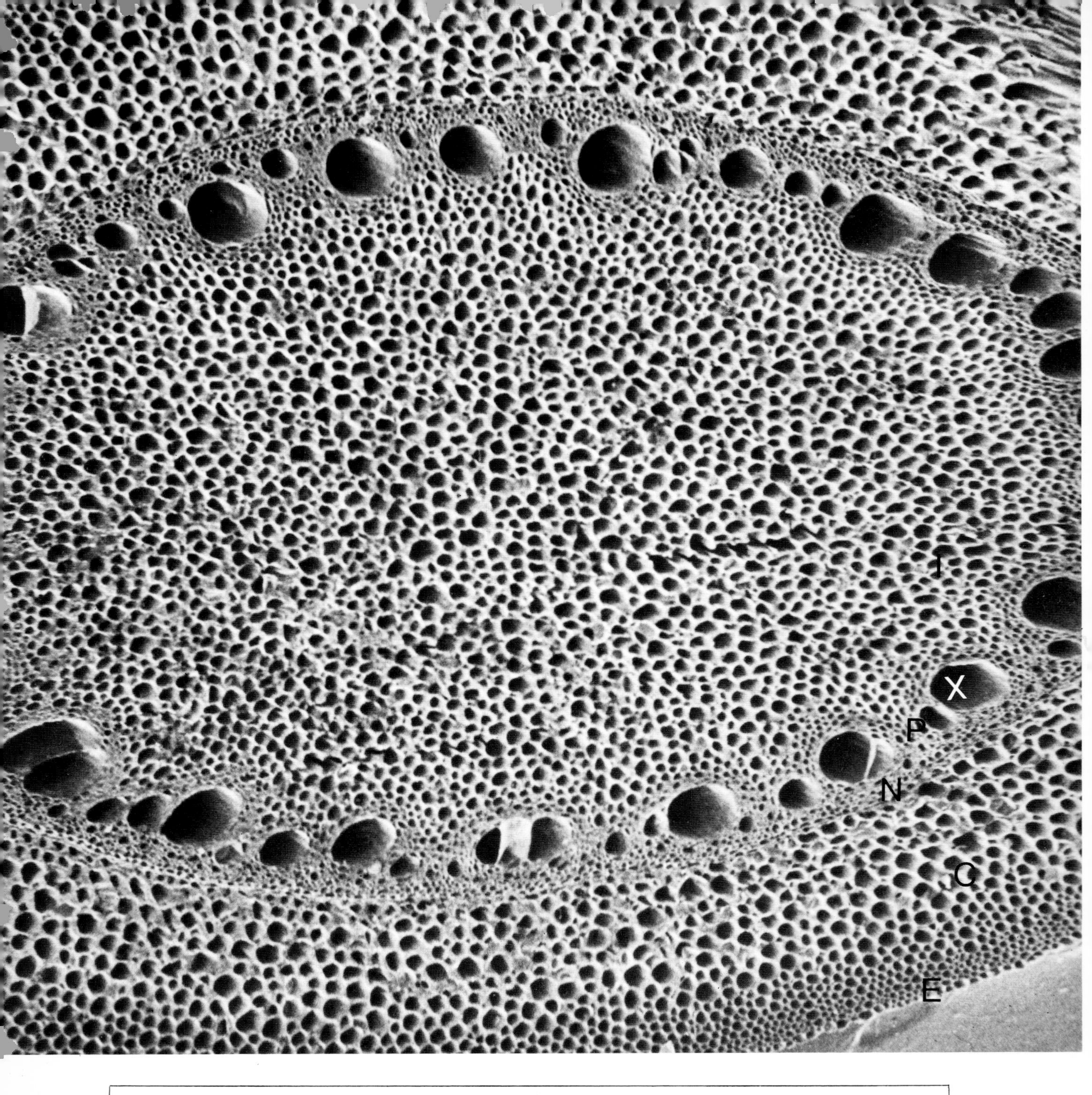

Plate 58 x90 — Xy!em

Transverse view of a prop root of maize. The prop root of maize is an adventitious root, that is it arises from the stem, which is above the ground and can be seen at the base of most maize plants. Numerous specialised types of cells make up the root and these include the epidermis (E), cortex (C), endodermis (N), phloem (P), xylem (X) and pith (I). The endodermis is important because it is a one-cell-thick ring of cells with a thickened band, the casparian strip, round the radial and transverse walls. This strip is made of a waxy material, suberin, which causes water and nutrients in solution to pass through a membrane and into the cell in order to enter the xylem vessels. Consequently the characteristics of the cell membrane can influence ion and water uptake by the plant.

Plate 59 x720 | Xylem

Xylem vessels in a prop root of maize. The roots of monocotyledons rarely have secondary growth and as can be seen in this photograph the internal surface is smooth. The markings running down the internal surface of the xylem vessels indicate the junction between the cells that surround the vessel. It is xylem elements such as these that provide a link between the water taken into the root and the water in the cells of the leaf.

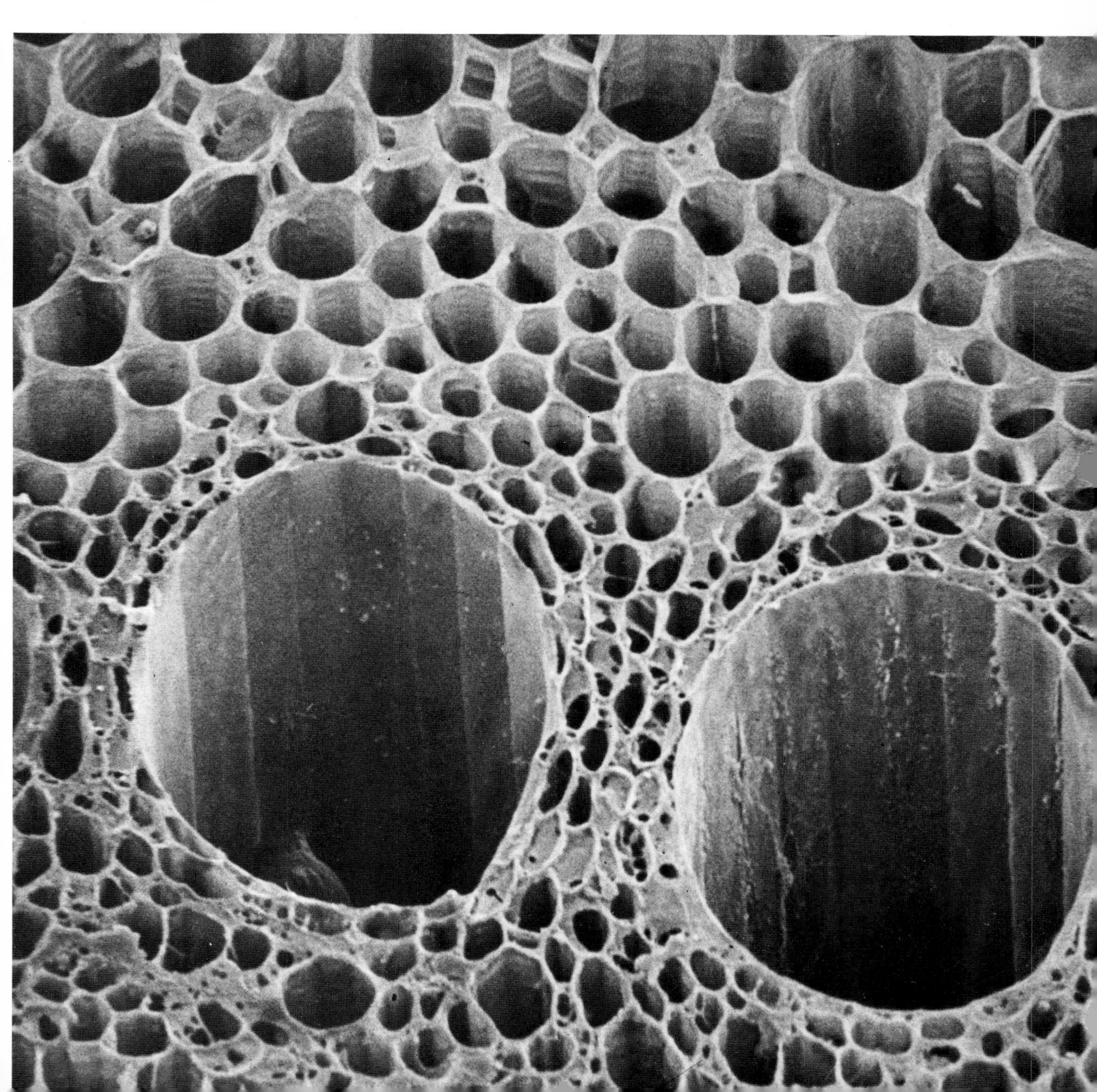

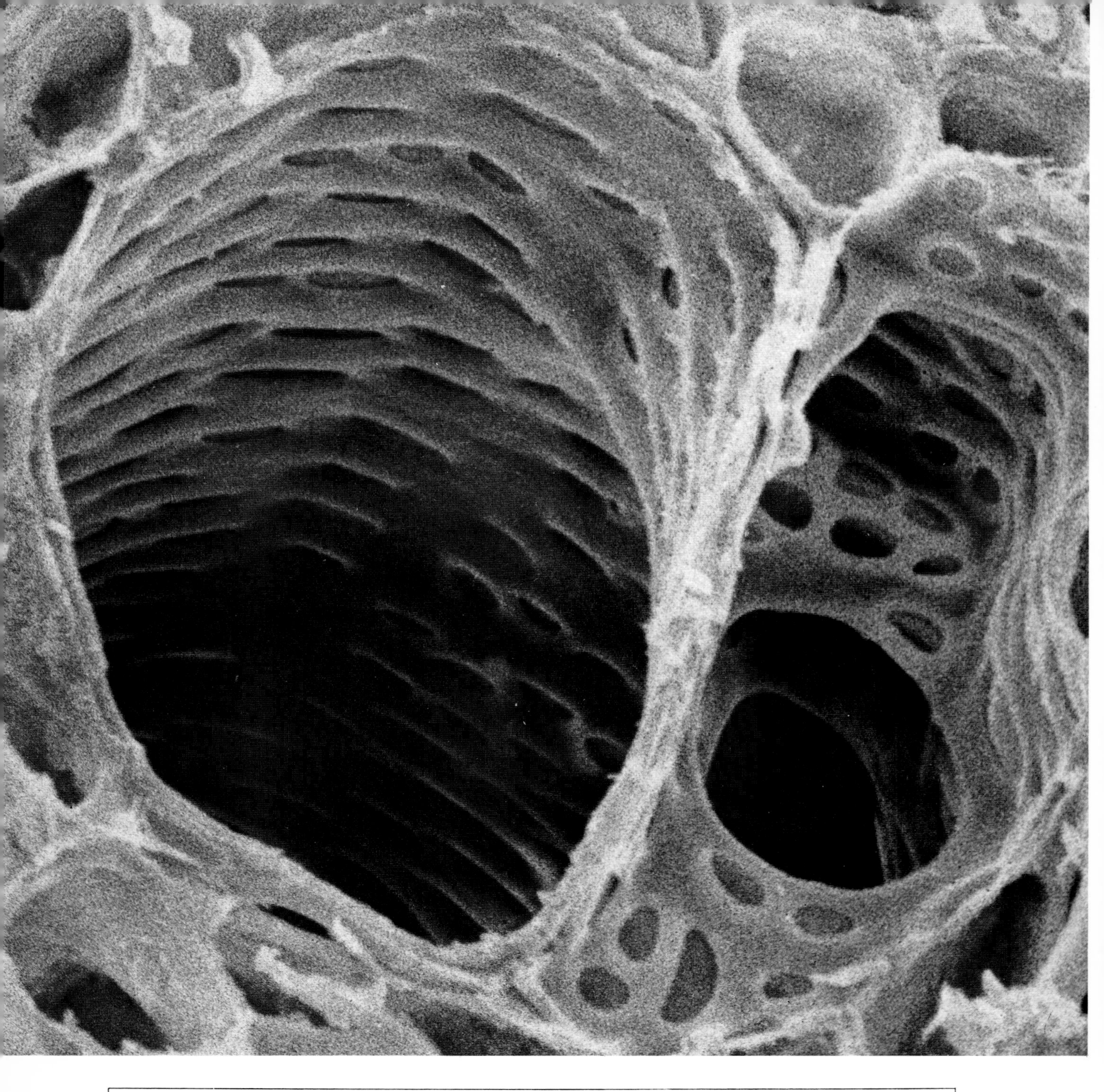

Plate 60 x2700 — Xylem

Xylem vessel in a cucumber root. Secondary wall thickenings are present and the xylem wall has a pitted appearance.

Plate 61 x1200	Xylem

Xylem vessels in a cucumber stem. The xylem vessels of the root link with the xylem vessels of the stem and as shown in this photograph the appearance of the cell wall in the stem is similar to the root.

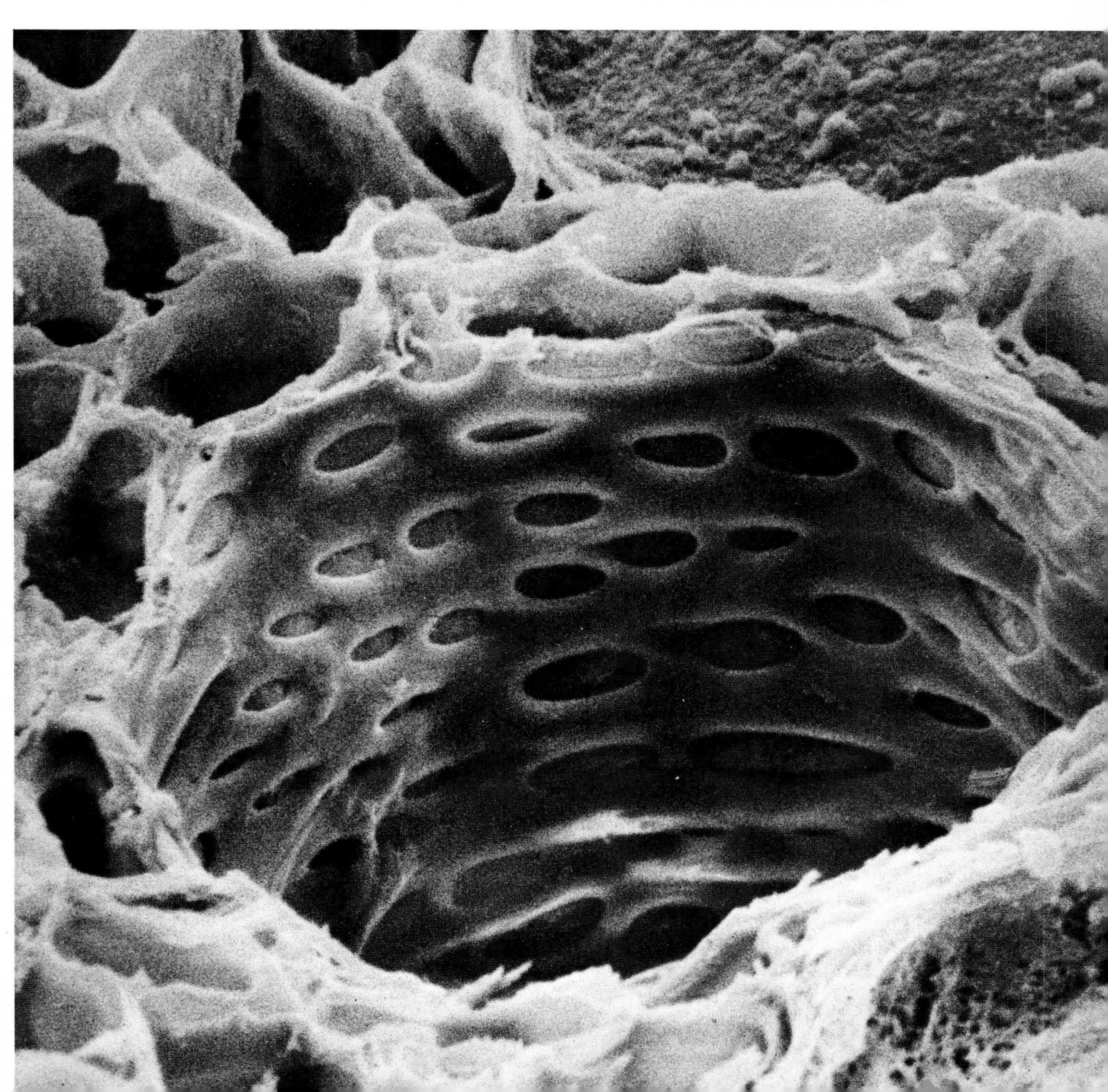

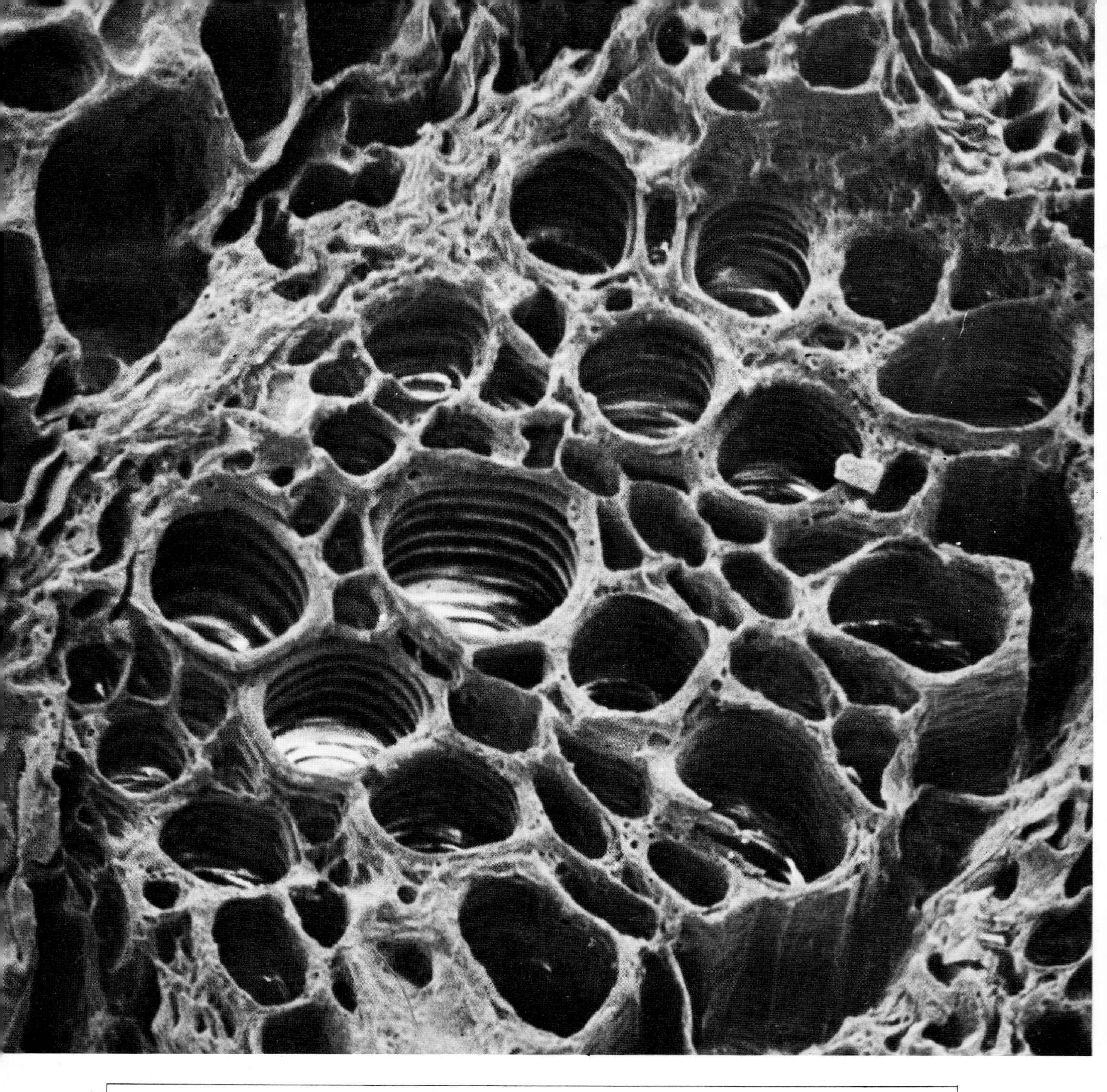

Plate 62 x4400 **Xylem**

Xylem vessels in a cotton *(Gossypium hirsutum)* **petiole.** Leaves of some plants such as cotton are large and provide an extensive area over which transpiration of water can occur. To prevent water stress during conditions inducive to high transpiration rates, it is necessary to have a plentiful supply of water to the roots and to have a large xylem system. If there were only a few xylem vessels or if they were too narrow they could restrict the supply of water to the leaves. In a cotton petiole there are a large number of xylem elements although the size and number of the elements depends on the area of leaf they supply with water.

Plate 63 x440 **Xylem**

Xylem vessels in a cucumber stem. Previous examples of xylem have shown that the internal cell wall surface may be smooth or pitted but there are several differences in the appearance of the wall that are related to the age of this tissue. The cell initially consists of primary cell walls but as increasing amounts of secondary tissue is laid down it progressively obscures the primary wall. In the earlier xylem elements the secondary wall may occur as rings while in the later stages the surface has a pitted appearance. There are several stages between these two types, some of which will be shown in later photographs. In this photograph of a cucumber stem there are the annular and spiral types on the left and the reticulate types on the right.

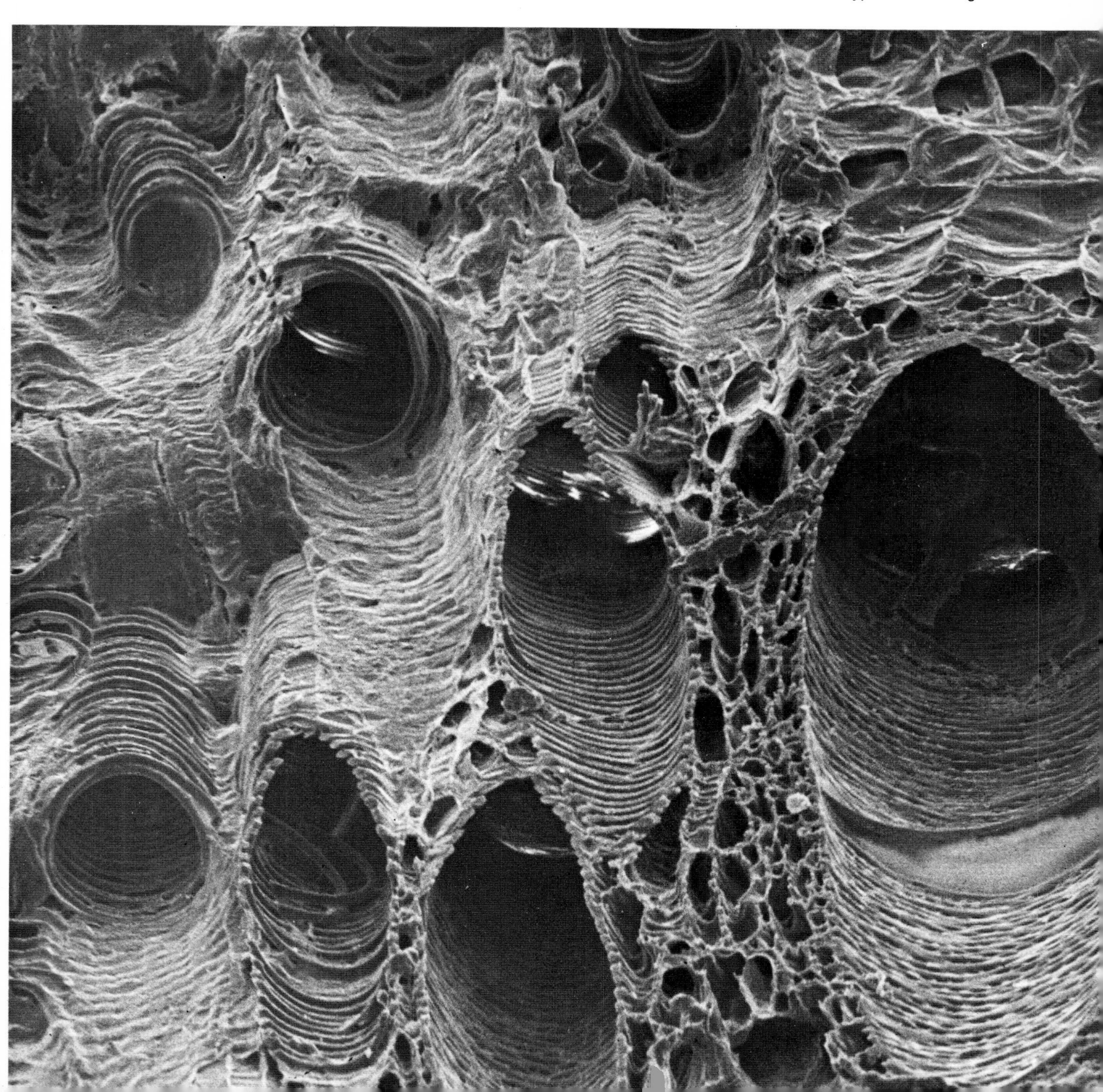

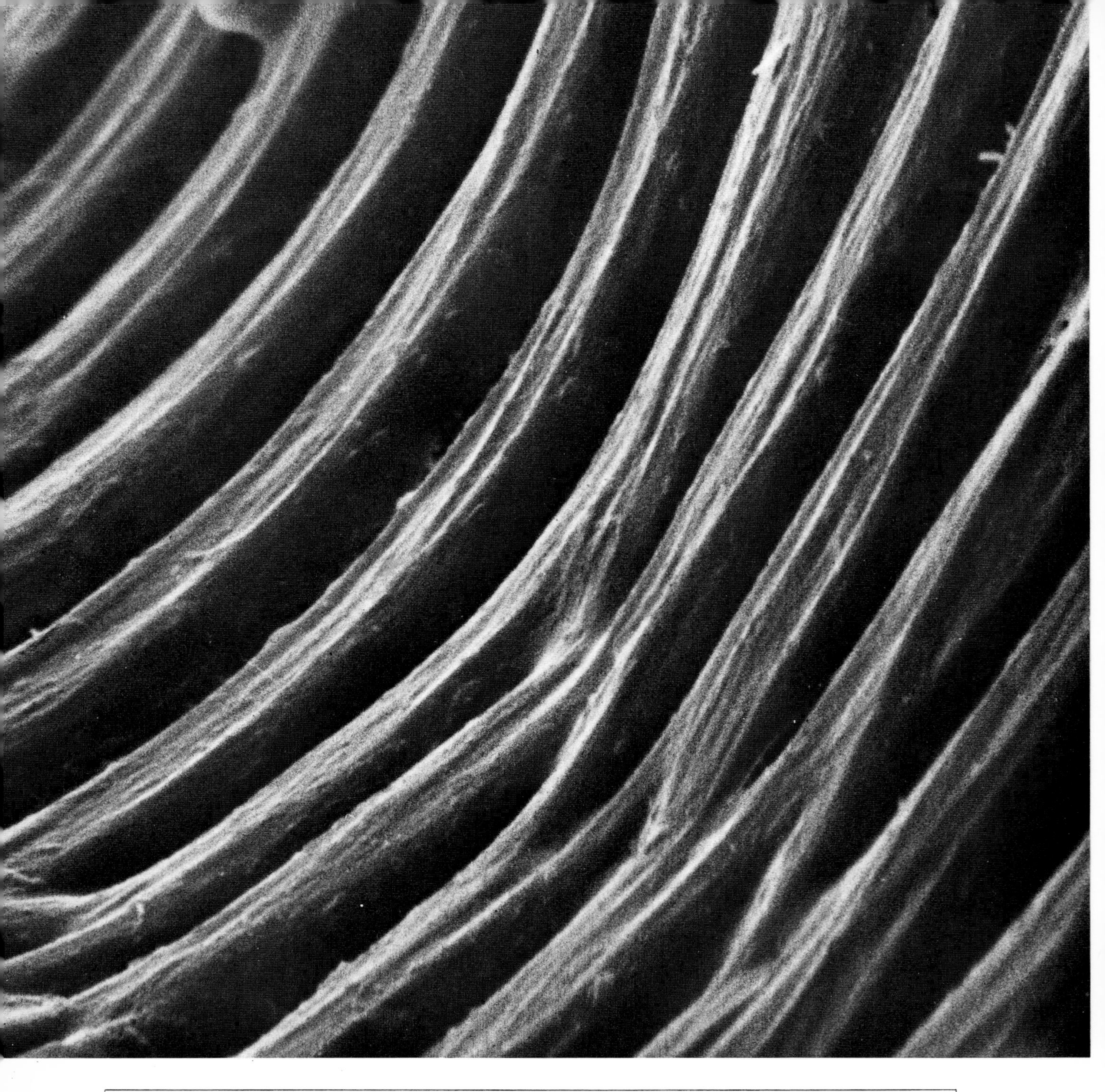

Plate 64 x7000	Xylem

Scalariform secondary thickenings in a xylem vessel from a cucumber stem. In this example the single bands of thickening are interconnected and termed scalariform thickenings. There does not seem to be any physiological significance for the variation in the organisation of the cellulose in the xylem cell walls.

Plate 65 x1600 — Xylem

Spiral secondary thickenings in a xylem vessel from a cucumber stem. The thin primary wall in these vessels has been broken in places between the spiral secondary thickenings of the secondary wall. The nature of the spiral thickenings can be seen very clearly in the cut ends of the vessel on the right of the photograph.

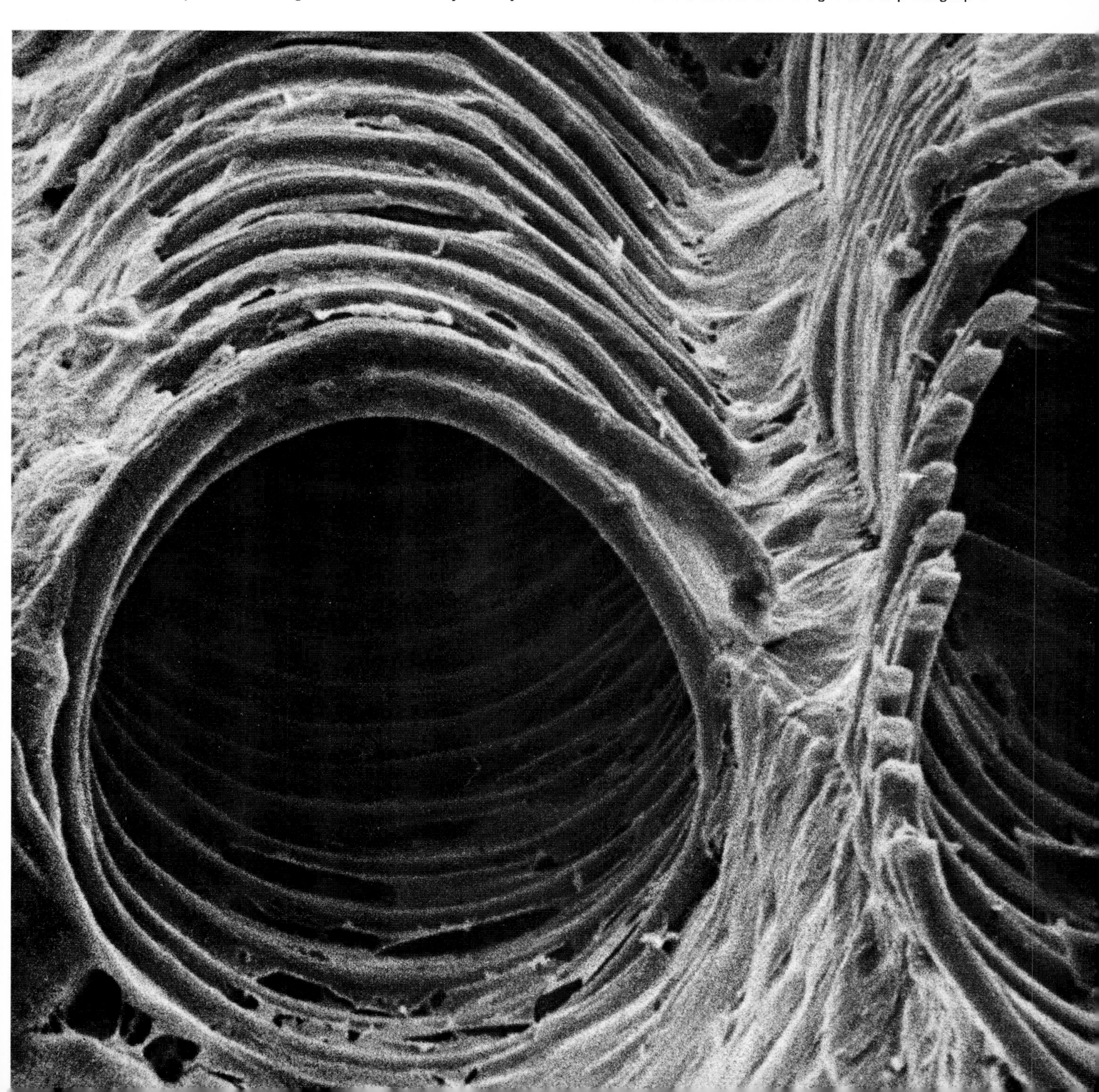

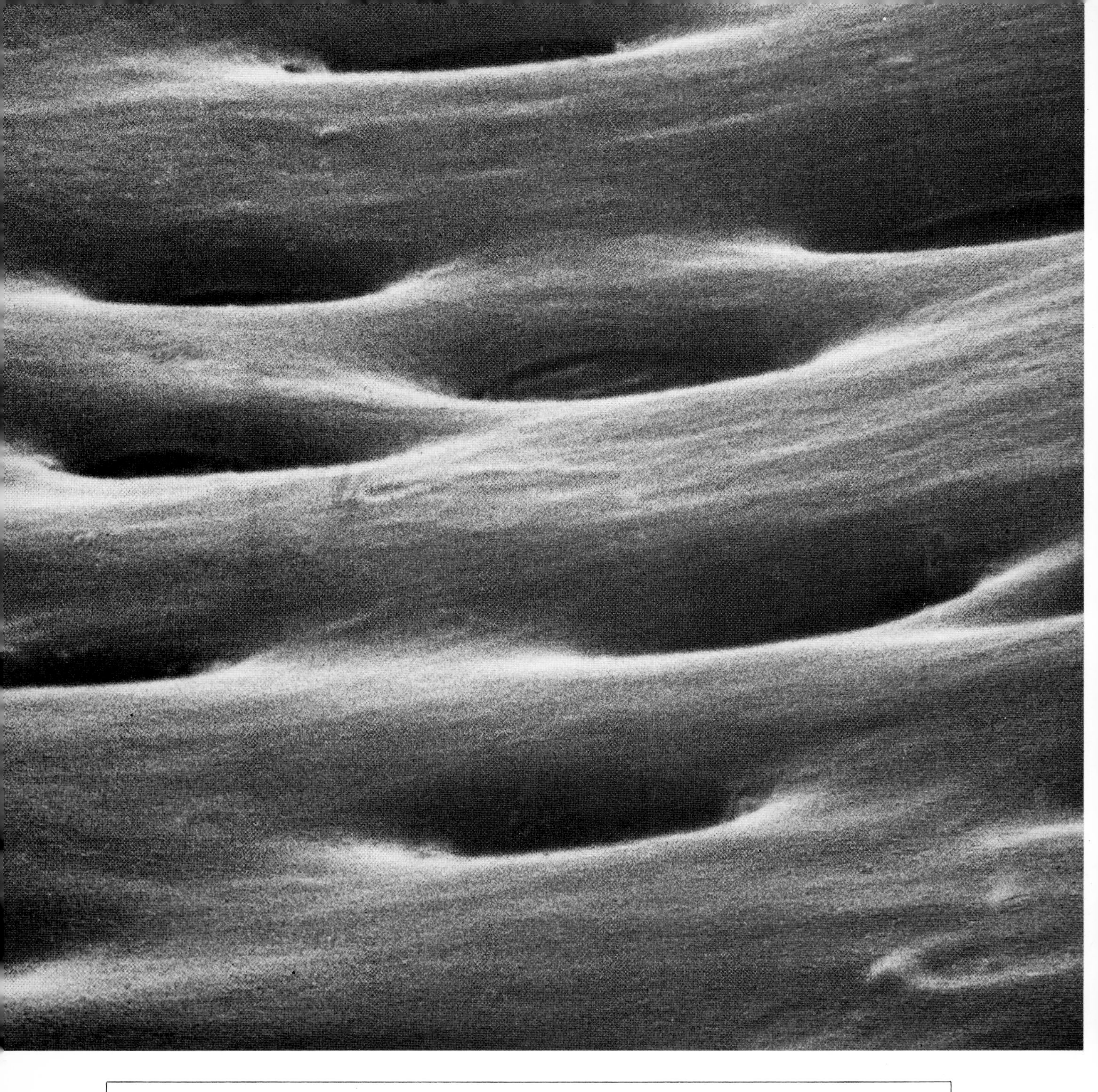

Plate 66 x12000 Xylem

Reticulate secondary thickenings in a xylem vessel from a cucumber stem. By progressive filling in of the spaces between the rings of secondary thickening the wall appears net-like. This view taken at a low angle to the wall emphasises the presence of the rings of thickening.

Plate 67 x6000 Xylem

Reticulate secondary thickenings in a xylem vessel from a cucumber stem. This photograph was taken directly above the inner cell wall surface of the xylem vessel and illustrates the thickenings that give a net-like appearance to the wall.

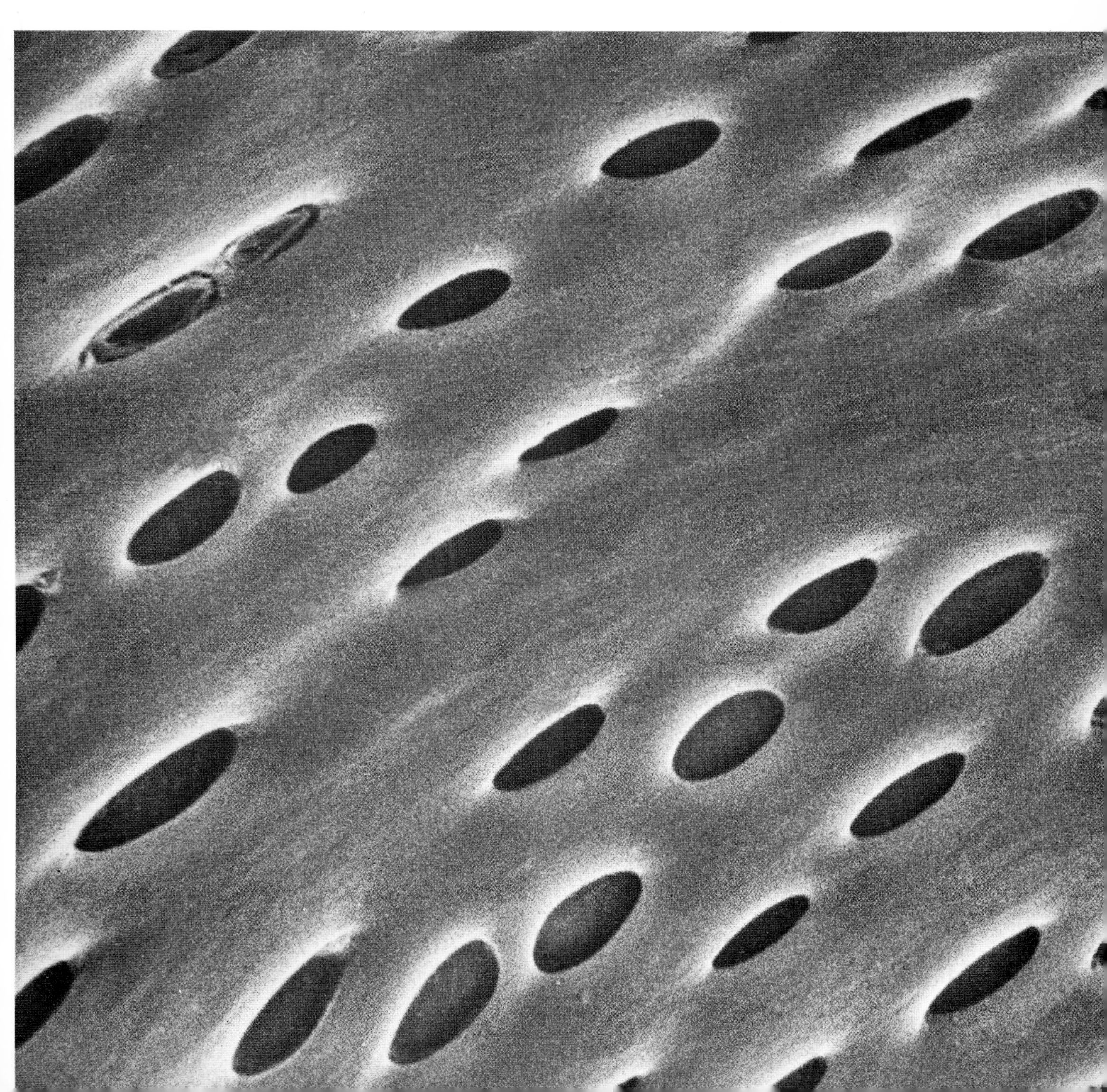

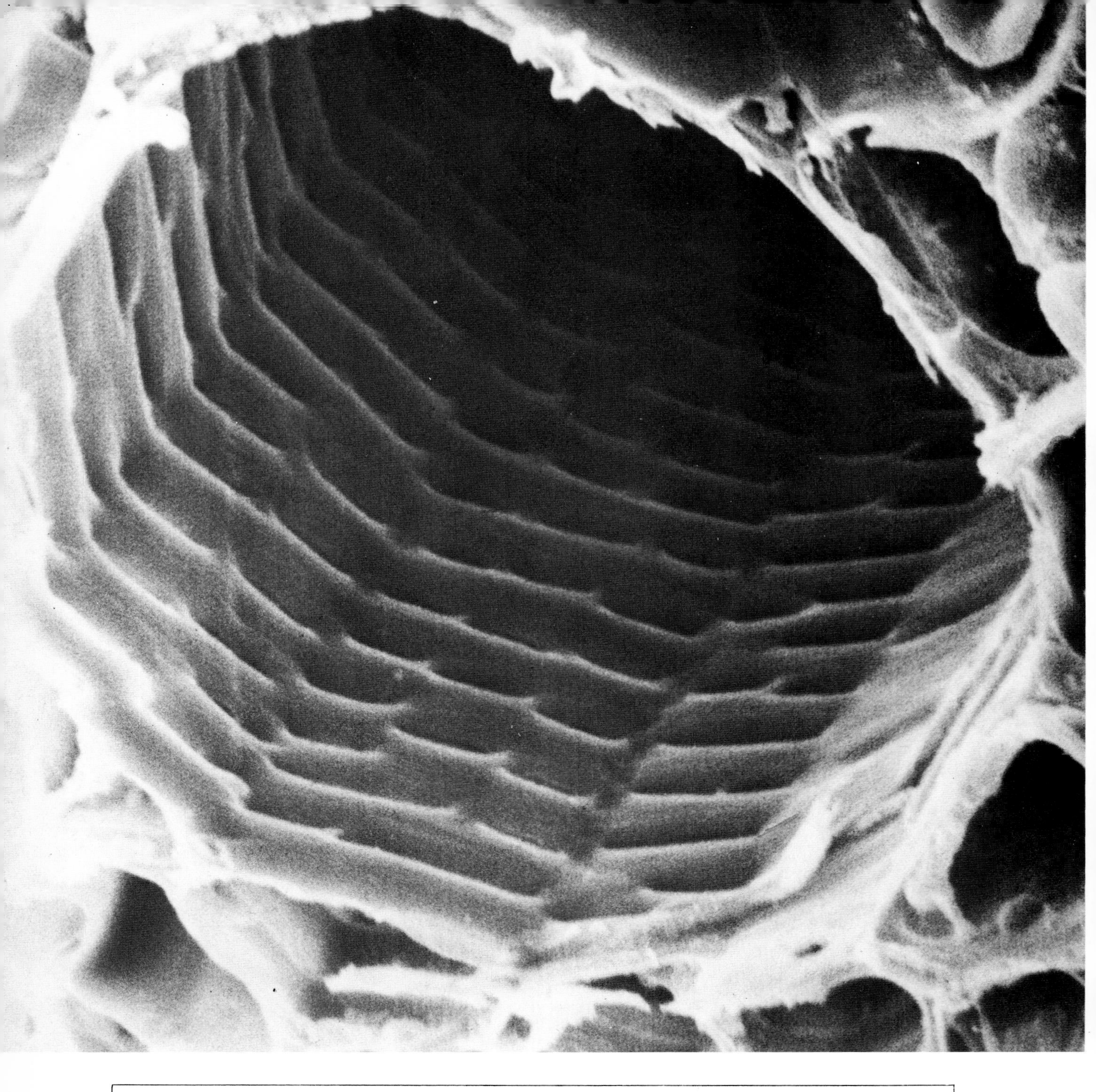

Plate 68 x4200 **Xylem**

Reticulate secondary thickenings in a xylem vessel from a maize leaf. The xylem vessels that have been shown in this series of photographs have been interpreted as being primarily involved in the conduction of water throughout the plant. These elements will also contribute to the rigidity of the plant although most of the support in larger plants is provided by the fibres.

The Apical Meristem

Plate 69 x200	The Apical Meristem

The apical meristem of Arawa wheat at the late vegetative stage. Most of the structures that have been illustrated so far in this book have been concerned with the vegetative aspects of growth. In particular, attention has been directed to the leaf because of its importance to photosynthesis. One of the primary activities of the apical meristem in the vegetative stage is to initiate new leaves. The tip of the apex consists of apical initials, which divide to form the shoot, and the leaves are initiated lower down on the side of the apex. The leaves first appear as a ridge but with successive cell divisions and expansions, the size of the ridge increases and takes on the more familiar shape of a leaf.

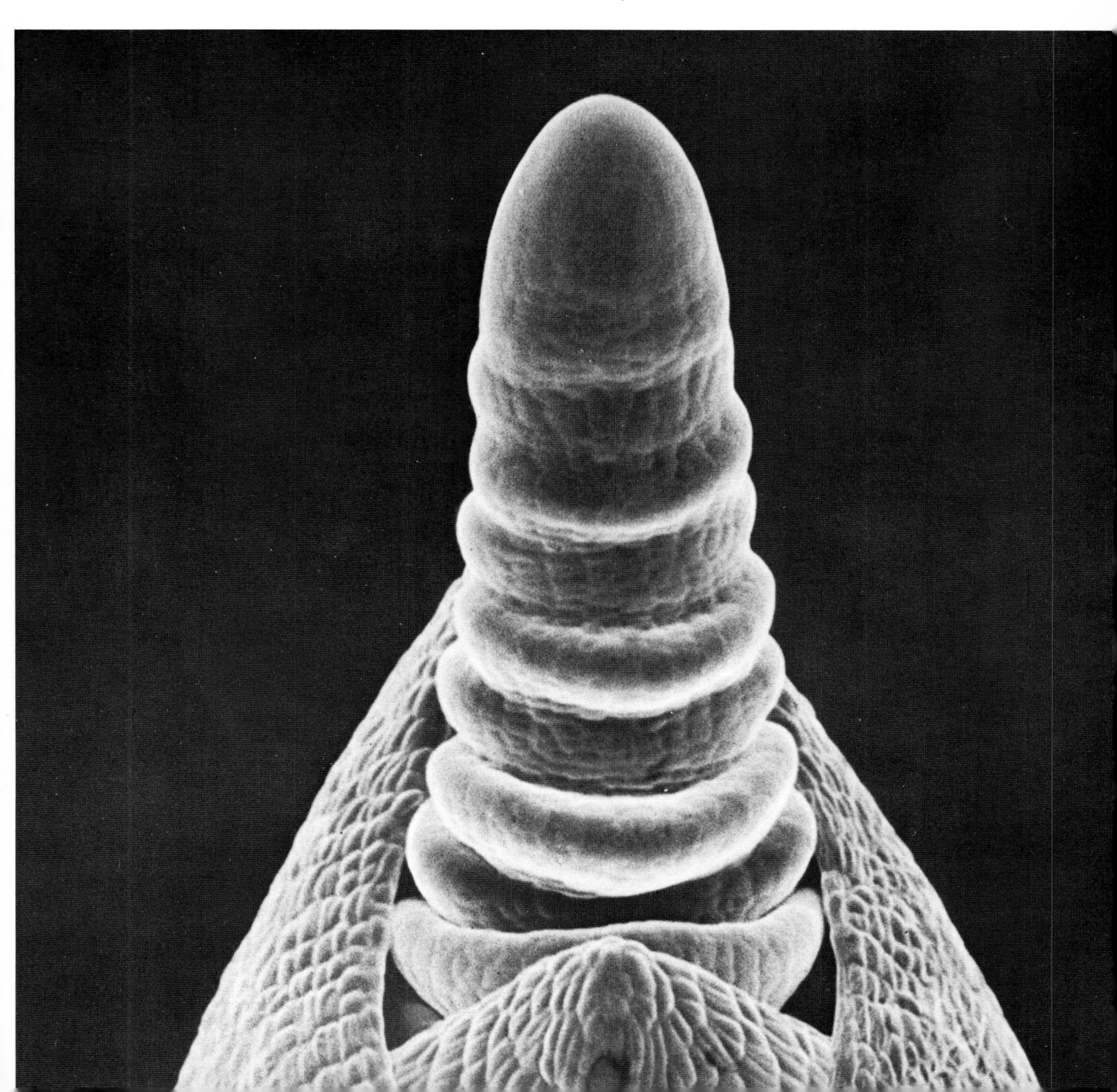

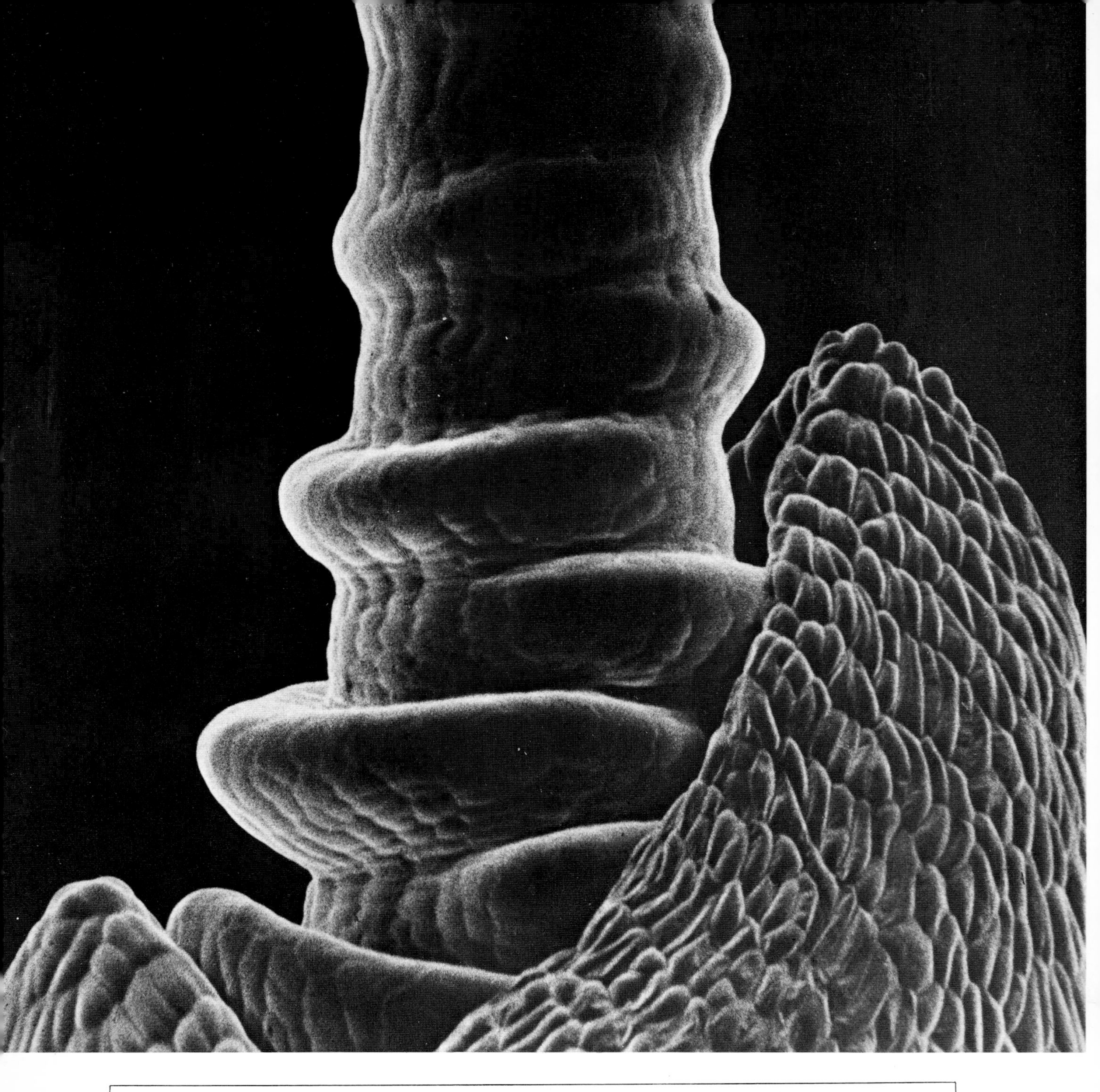

Plate 70 x380 — The Apical Meristem

The apical meristem of Arawa wheat at the late vegetative stage. To obtain this view of the apical meristem it was necessary to dissect away several fully and partially expanded leaves which normally surround and protect the apex. The individual cells of the leaf can be seen on the surface of one of the leaflets partially enclosing the apex. The apical meristem in this example is elongated, which indicates that the switch from vegetative to reproductive growth has been initiated. The signal for this switch is received by the leaves as a result of a change in environmental factors—such as day length. The first visible signs, however, occur in the apex, which elongates, and the later-informed single ridges develop into a double ridged structure due to the initiation of spikelet primordia in the axils of the leaf primordia.

Plate 71 x160 — The Apical Meristem

The apical meristem of Raven wheat in the reproductive stage. In the reproductive stage the single ridges on the apex are replaced by more complex structures which are collectively referred to as the spikelet. Apical meristems such as this one have been irreversably directed into reproductive growth and no more leaves will be initiated. The apical meristem at this stage has taken on the more familiar appearance of a spike, the central axis being the rachis and each branch the spikelet.

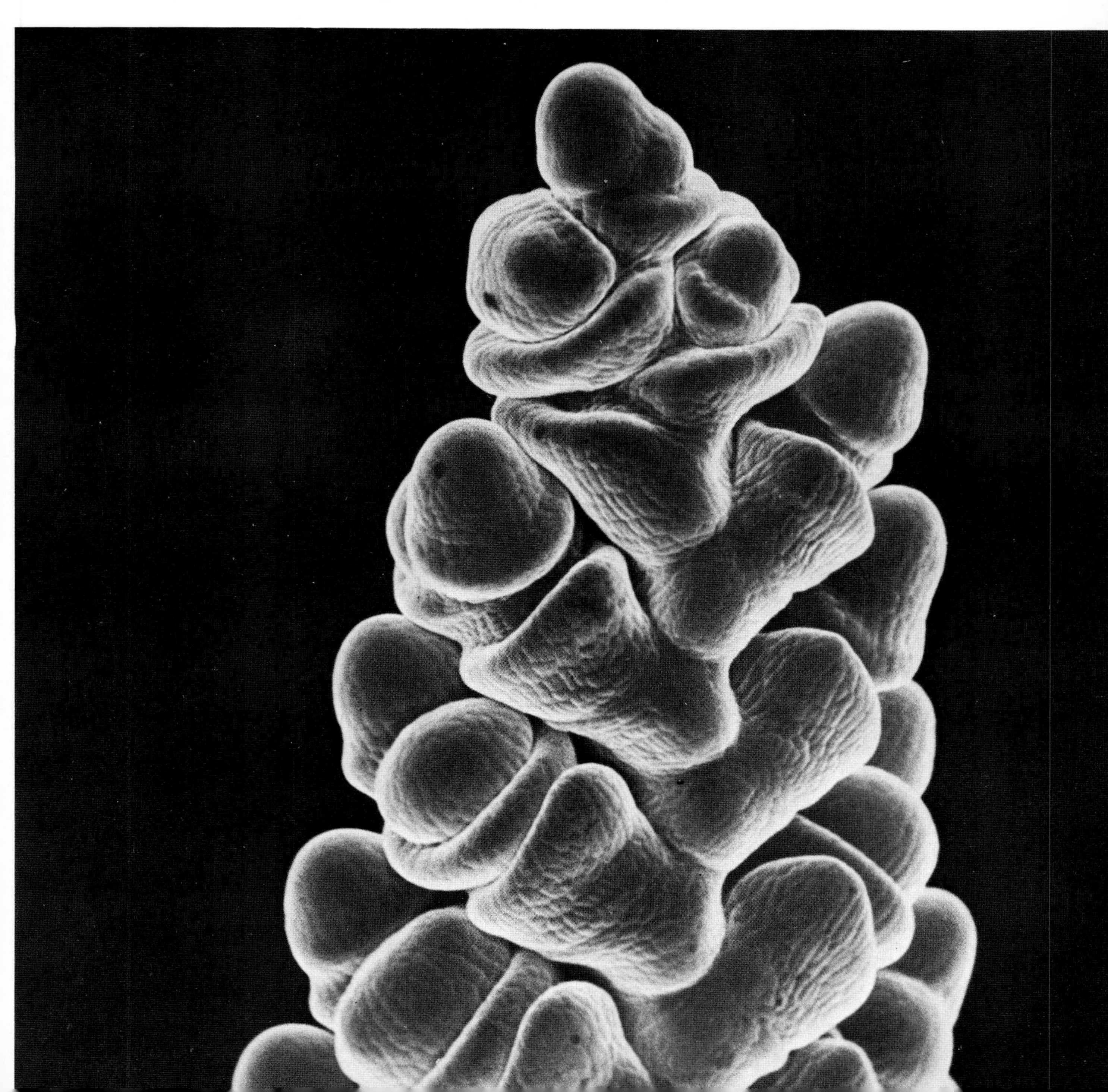

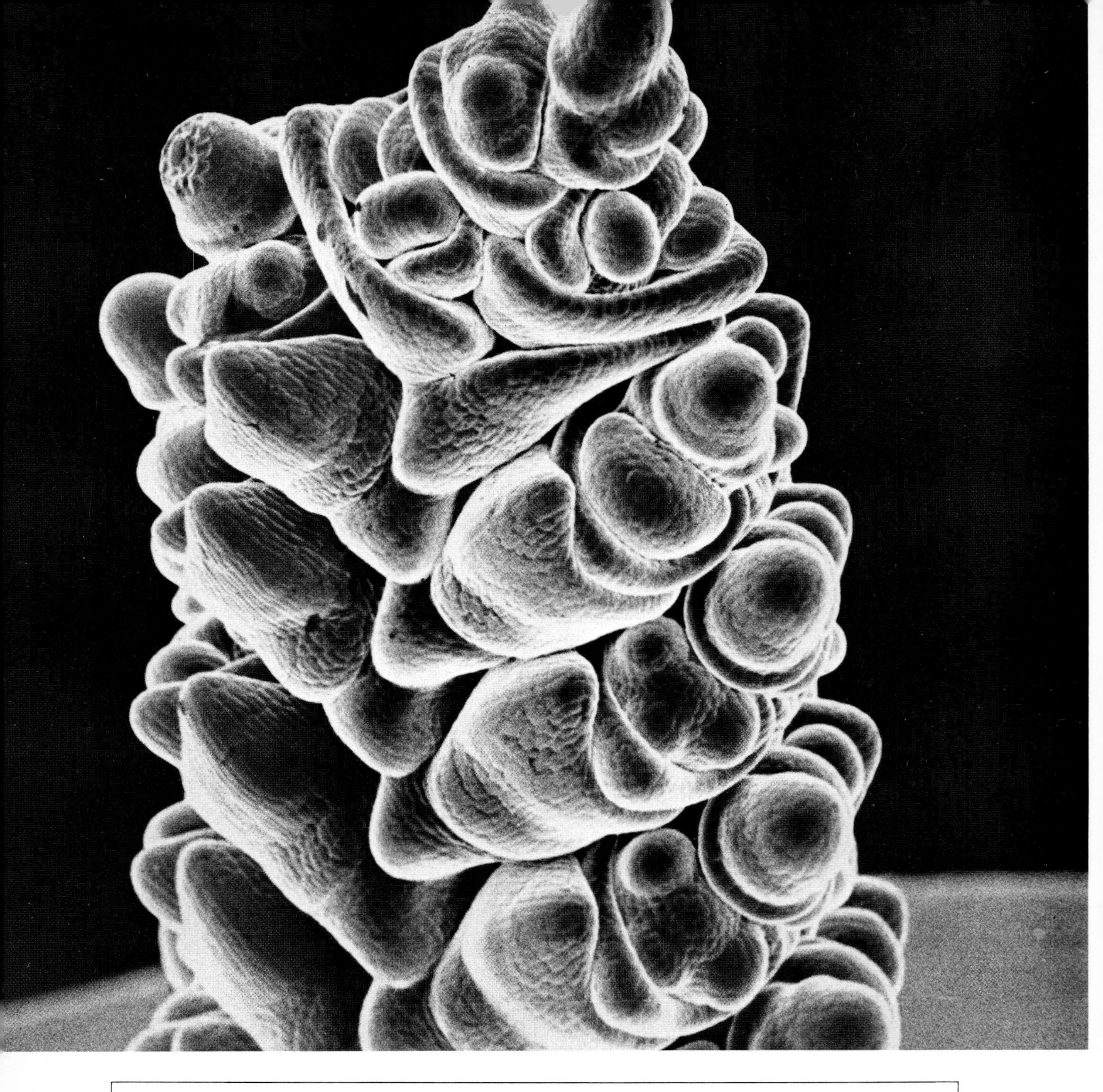

Plate 72 x190 — The Apical Meristem

The apical meristem of Raven wheat in the reproductive stage. With the continued development of the apex, prolific cell division results in the complex development within each spikelet. By this stage each spikelet can consist of up to six flowers, many of which do not produce seeds. The later formed spikelets are less well developed compared with earlier ones, as illustrated by successive spikelets down the apex shown in this photograph.

Plate 73 x352 | The Apical Meristem

The apical meristem of Raven wheat in the reproductive stage. At this advanced stage of reproductive development several structures are beginning to appear in each floret. Especially pronounced at this stage are the glume (G), lemma (L) and flower primordia (F) which contains the stamens (S) and carpels (C).

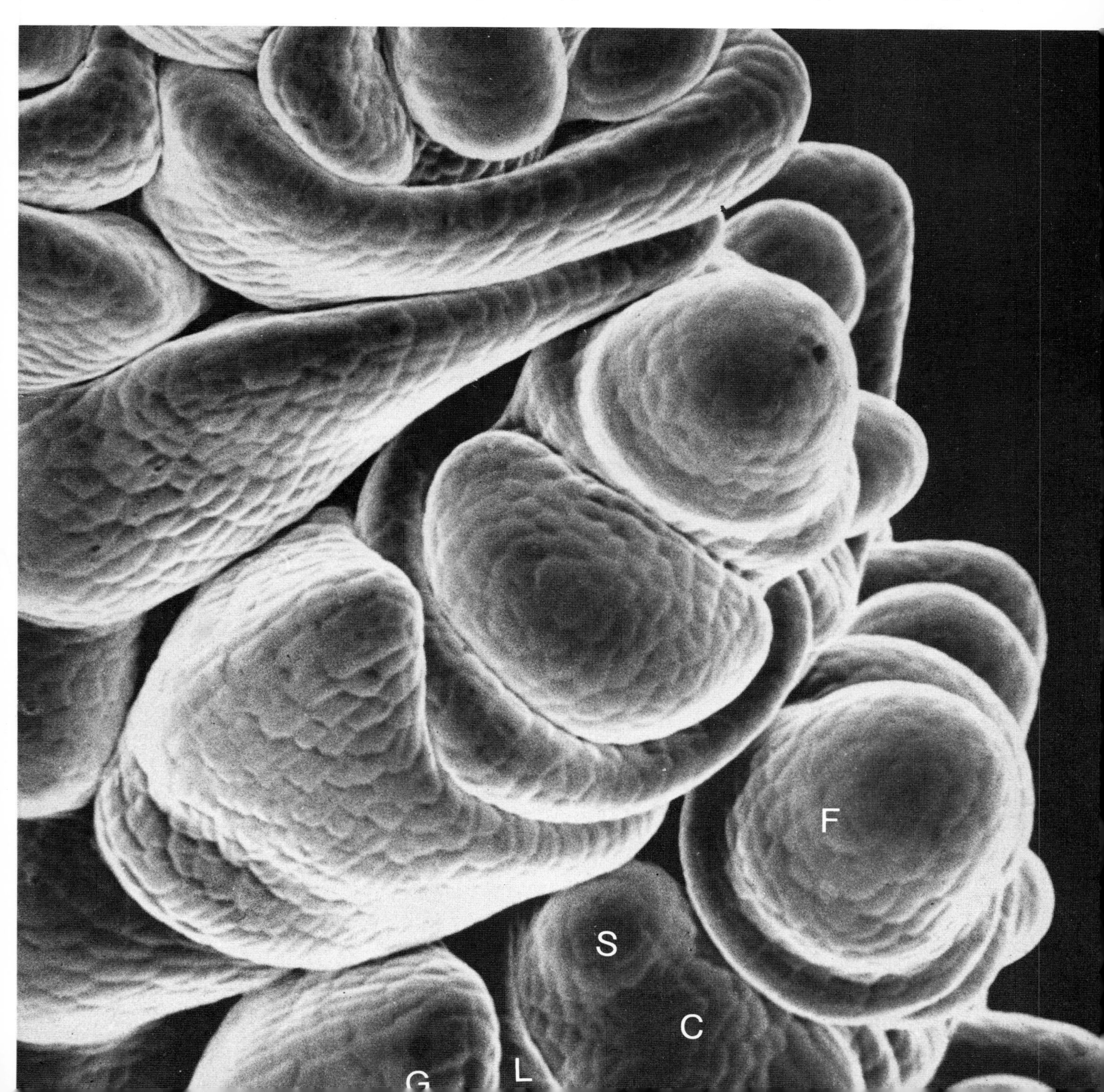

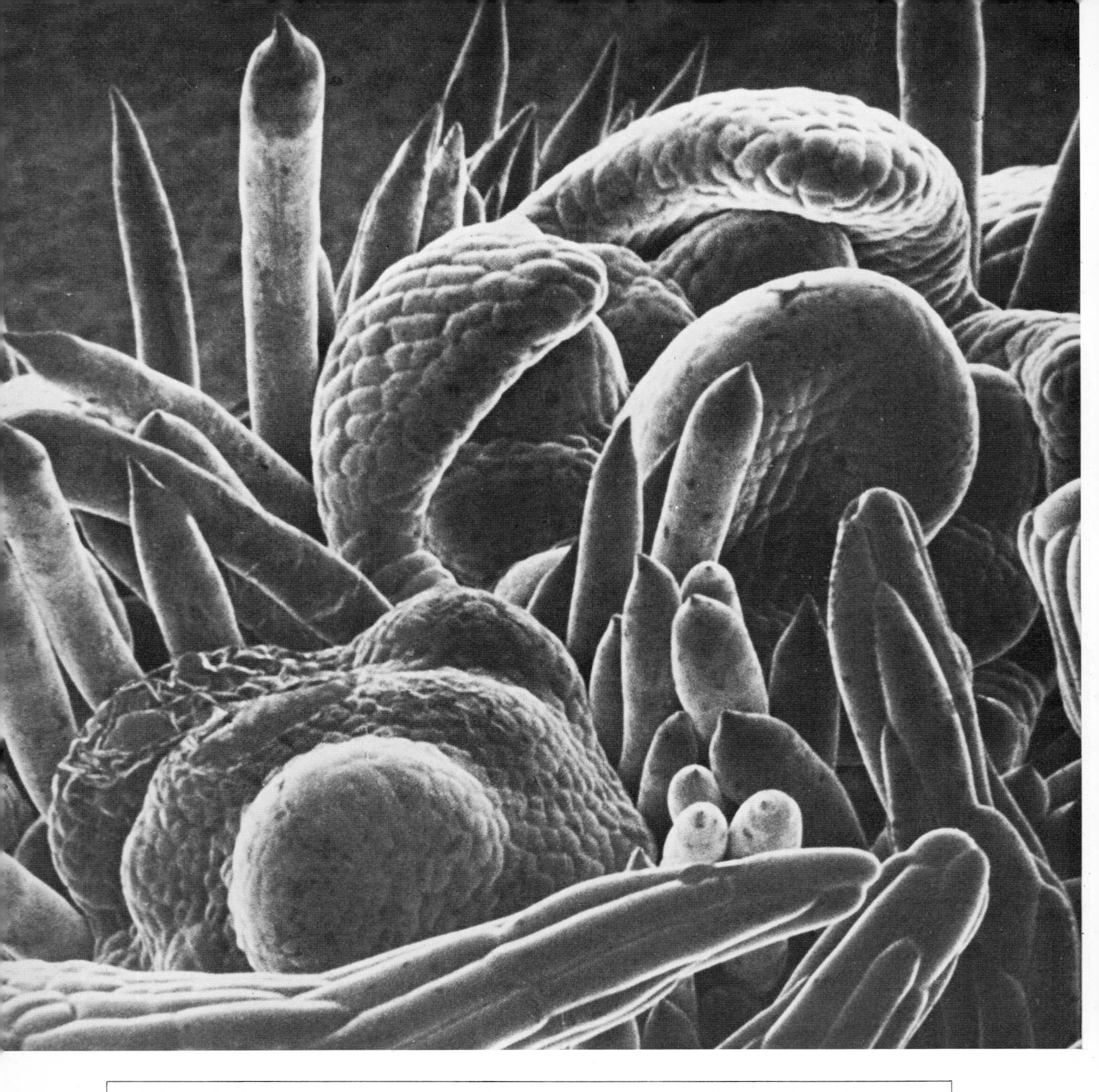

Plate 74 x150 The Apical Meristem

The floral apex of broad bean. The broad bean is a dicotyledon but the general appearance of the apex is similar to wheat. It is not possible at this early stage of flower development to distinguish between the numerous organs that make up a flower. However the order in which these organs will develop is defined; firstly the sepals, then the petals, followed by the stamens and finally the carpels.

Plate 75 x130 — The Flower

Flower of alyssum. The four major parts of a flower are illustrated in this photograph. There are the sterile organs which include the leaf-like sepals (S) and the colourful parts, the petals (P), and the reproductive parts, the stamens (St) and the carpels (C). The six stamens occur in a whorl round the carpels, with the anthers and stigma at about the same level which would be expected to result in rapid and efficient transfer of pollen from the anther to the stigma. This relationship between anthers and stigma does not hold in many other species, especially those that have attempted to prevent fertilisation within the same flower. Hence, for example, the anthers may be above or below the stigma, these two organs may develop at different times or they may be in separate flowers on the same or different plants.

Plate 76 x1100 — The Flower

Cells on the surface of a petal of alyssum *(Lobularia maritima).* A feature of most flowers other than those of the grass flower is the presence of petals. Extensive development of petals has occurred, which has resulted in the wide variation in colour, shape and size that are frequently seen. The photograph illustrates the nature of the epidermal cells of a petal which would be important in imparting an impression of texture to the flower.

Plate 77 x80 | The Flower

The flower of paspalum *(Paspalum sp.)*. Subsequent development in the paspalum spikelet leads to the growth of the lemma (L) and palea (P) which eventually enclose the floret. When the flower is mature the anthers (A) and stigma (S) are exposed to allow fertilisation.

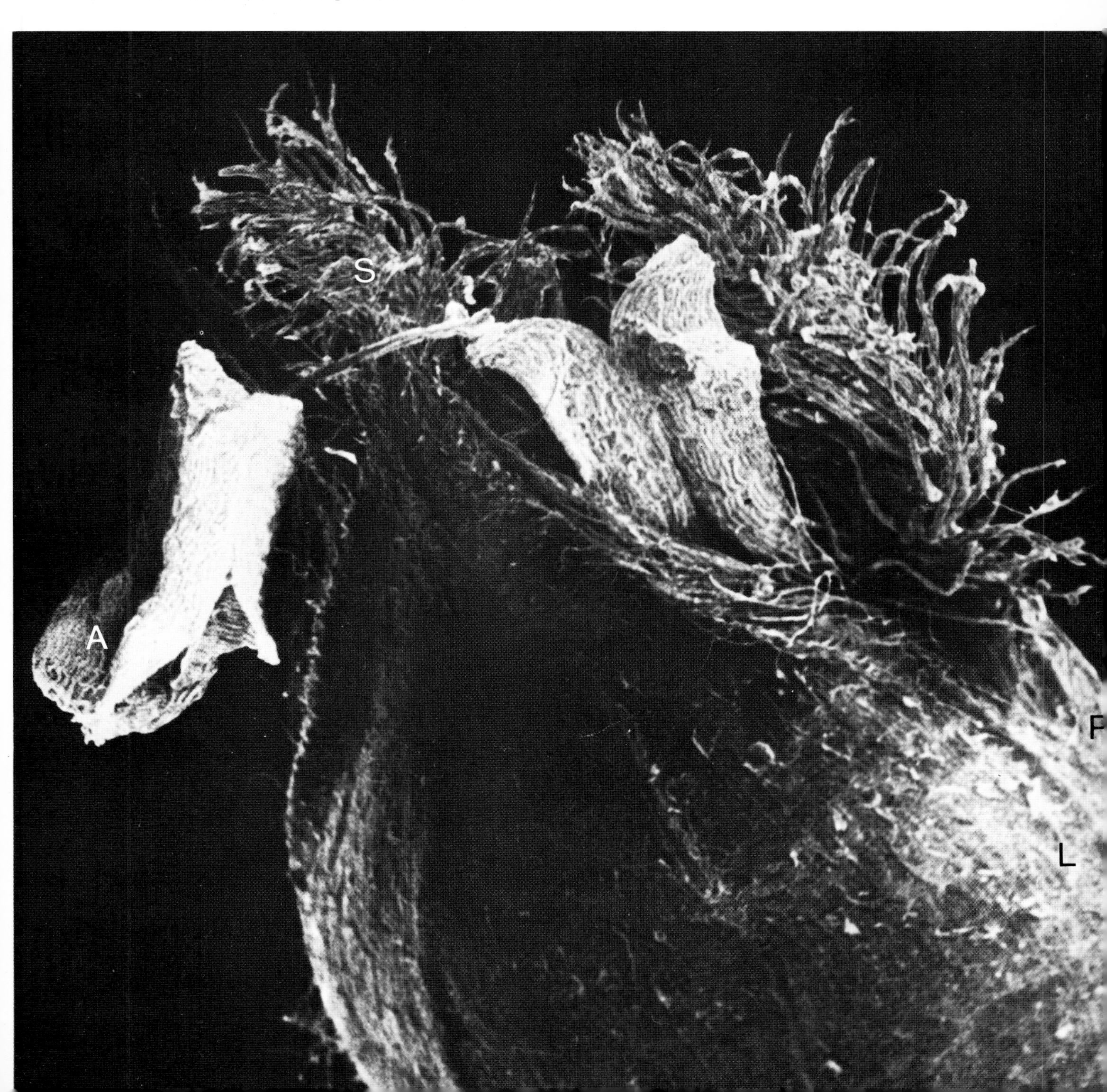

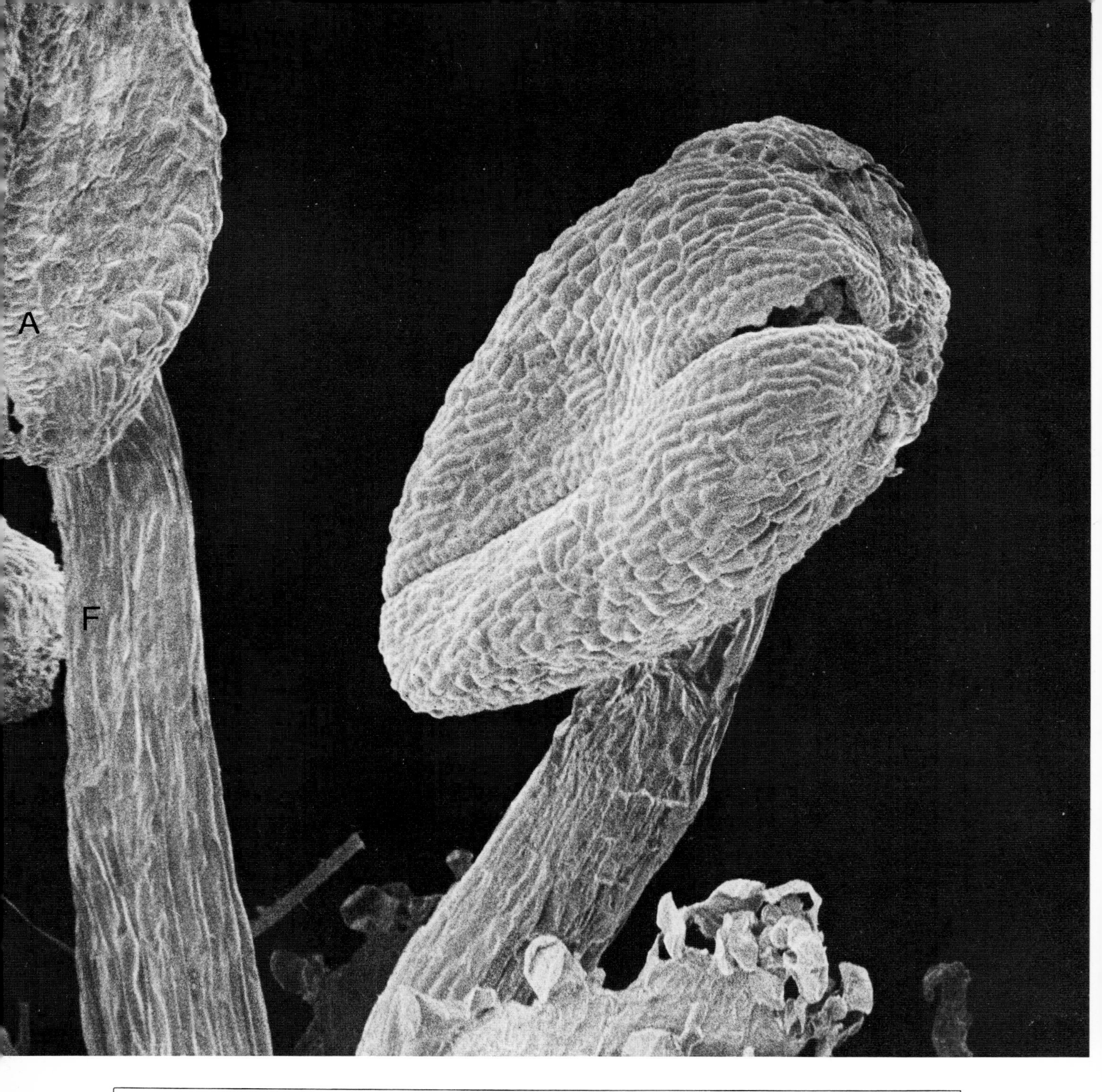

Plate 78 x50 The Flower

Anthers from a saltbush flower. The stamen consists of an anther (A) and filament (F) and stamens are the male parts of the flower. The pollen grains are the male gametophytes and they are initiated, develop and mature within the anther which has vascular connections to the rest of the plant through the filament. To release the pollen the anthers commonly dehisce along the line running along the anther.

Plate 79 x190 The Flower

Anthers, covered in pollen grains, from a cotton flower. In this example the anther has matured and split but many of the cotton pollen grains have remained attached to the anther.

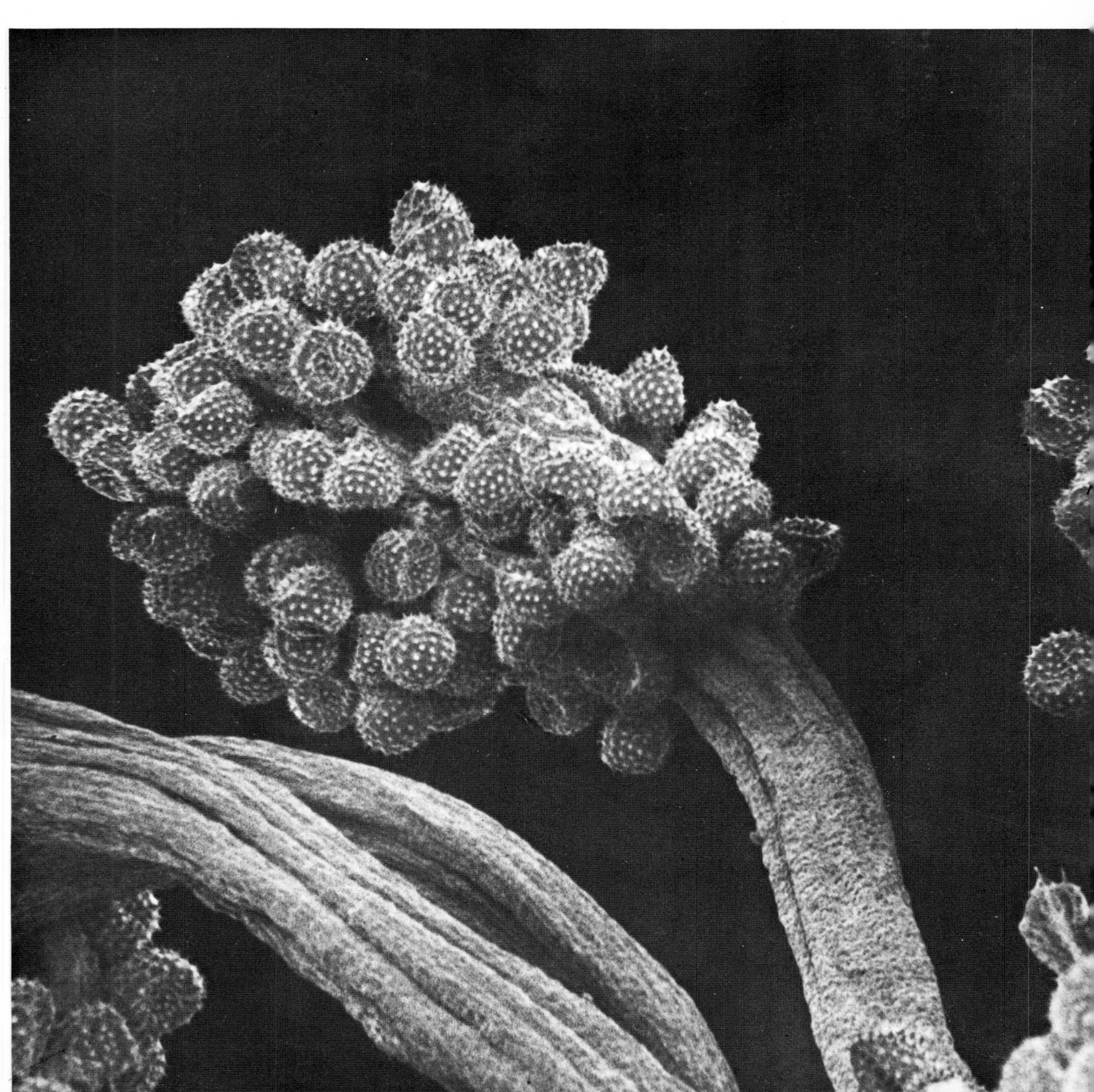

Plate 80 x160 — The Flower

Transverse view of an anther from a maize flower. The maize anther consists of a four loculed structure, connected through a single vascular element to the rest of the plant.

Plate 81 x200 The Flower

Transverse view of a locule from a maize anther. The anther is surrounded by the epidermal layer of cells, with a fibrous layer, the endothecium, inside. This photograph is taken at a late stage in development of the anther and shows the pollen grains neatly stacked within the locule. Normally another layer of cells, the tapetum, surrounds the pollen grains and provides nutrients which are apparently taken up through the pollen grain wall. The anthers open by a longitudinal split occurring in the cell wall, at a point between the locules where the epidermis is thin.

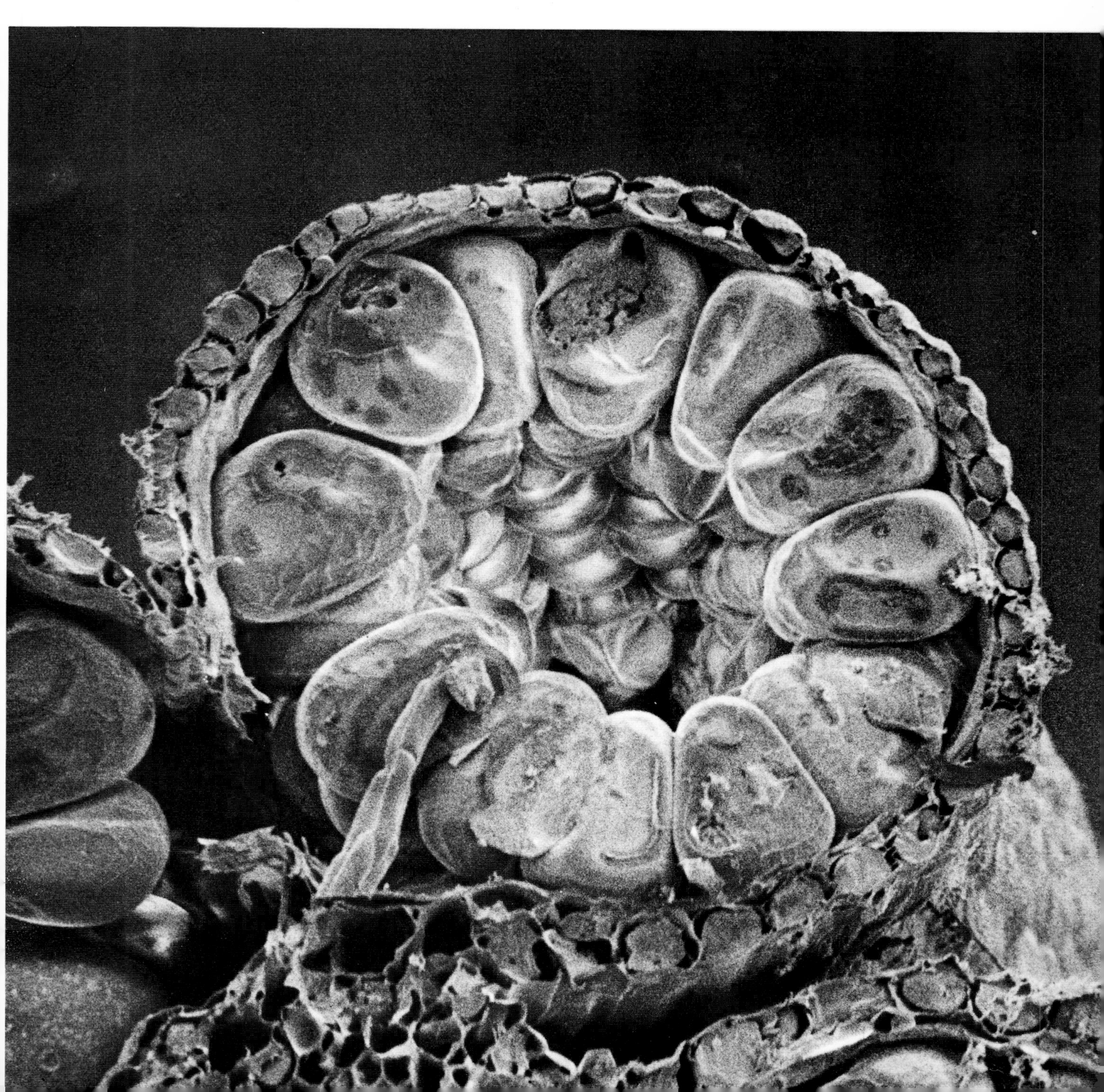

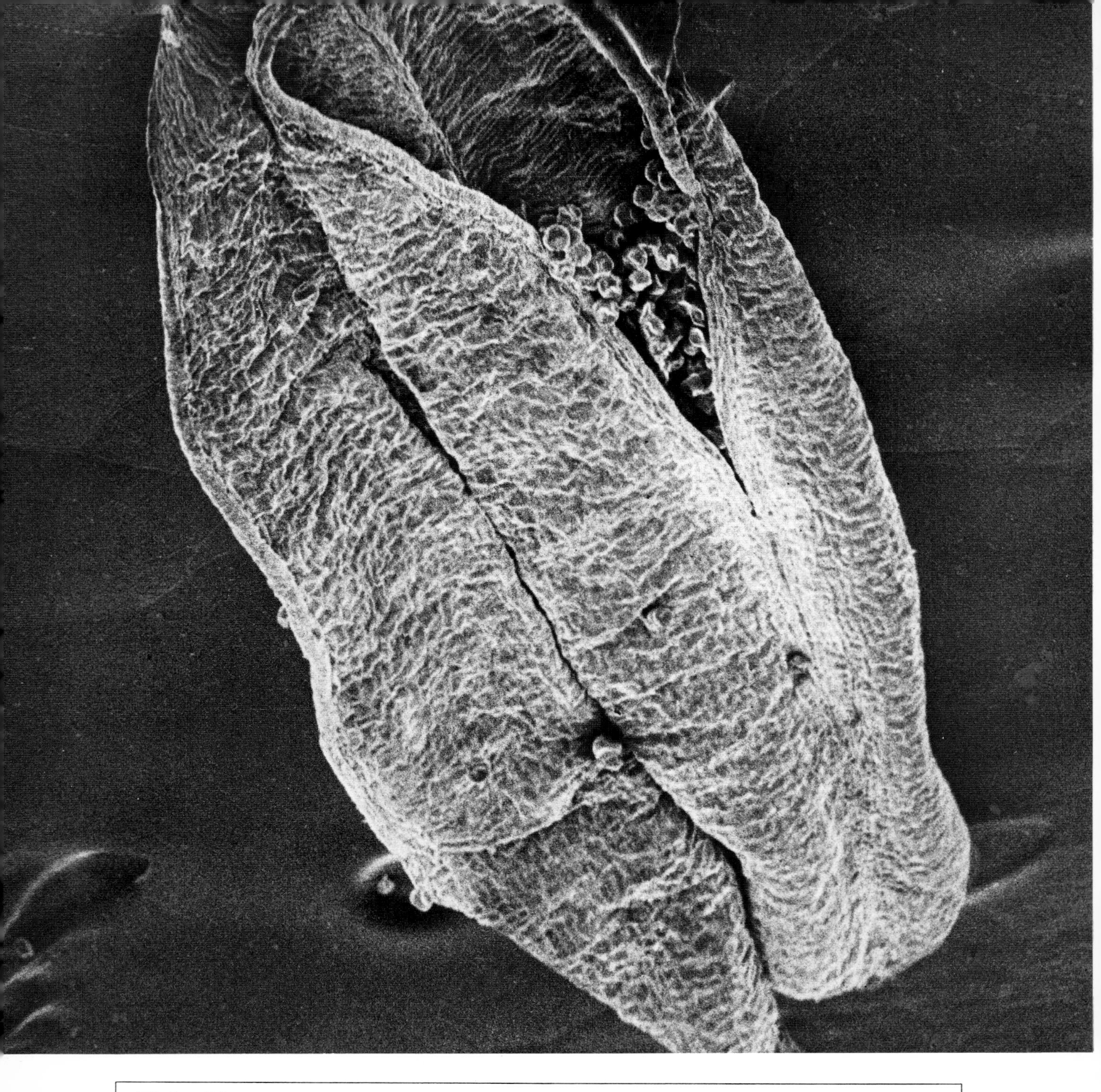

Plate 82 x420 | The Flower

Anther with pollen grains in wheat. The anther in this example has split and the pollen grains have been released. In this type of plant the pollen is dispersed by wind, but often other agents are involved, for example, insects, birds, mammals or water. Wheat is generally a self pollinating plant so that dispersal is over short distances.

Plate 83 x100	The Flower

Surface features of a wheat anther. The epidermal cells in the wheat anther are heavily cutinised as protection from dehydration.

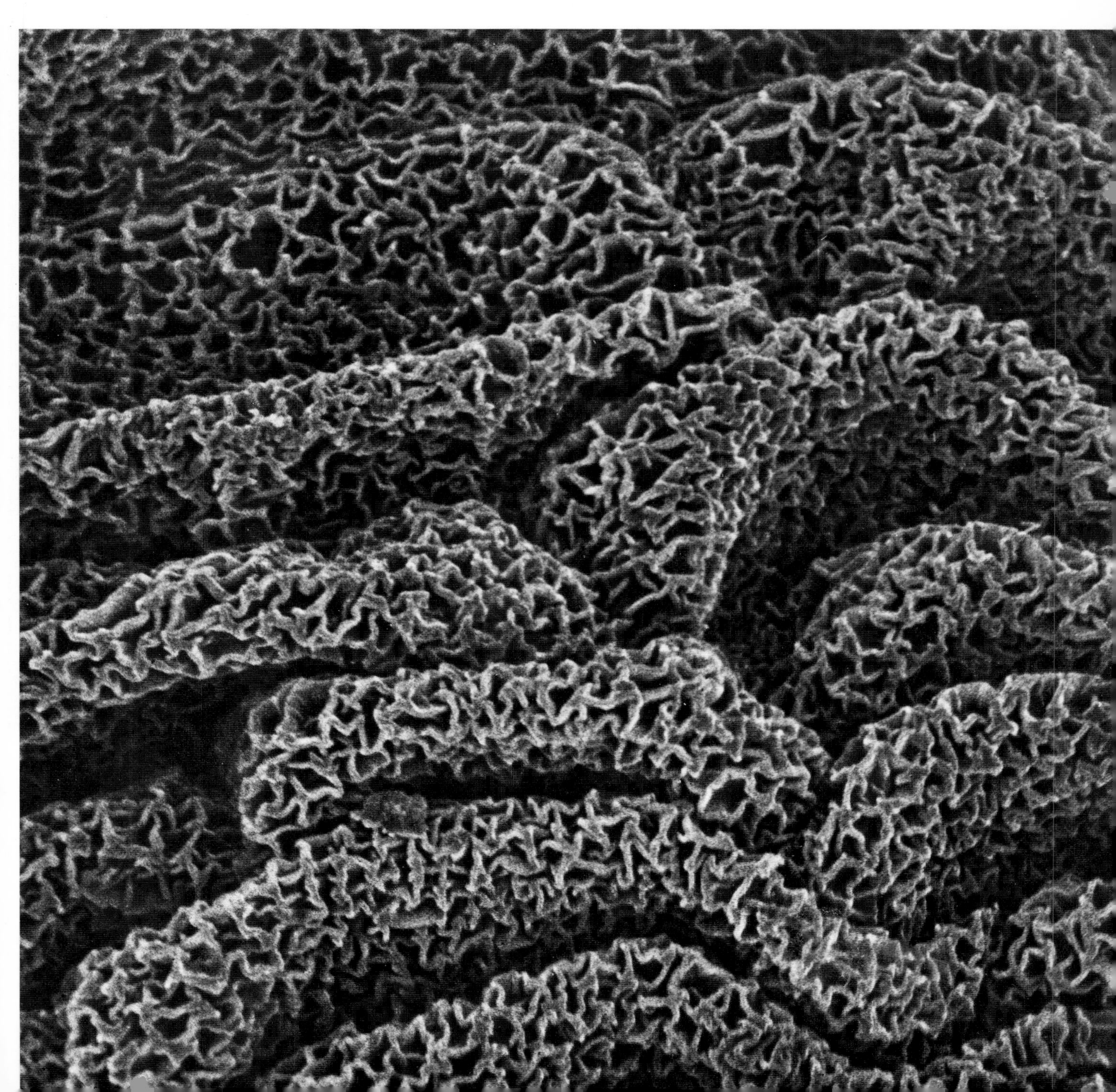

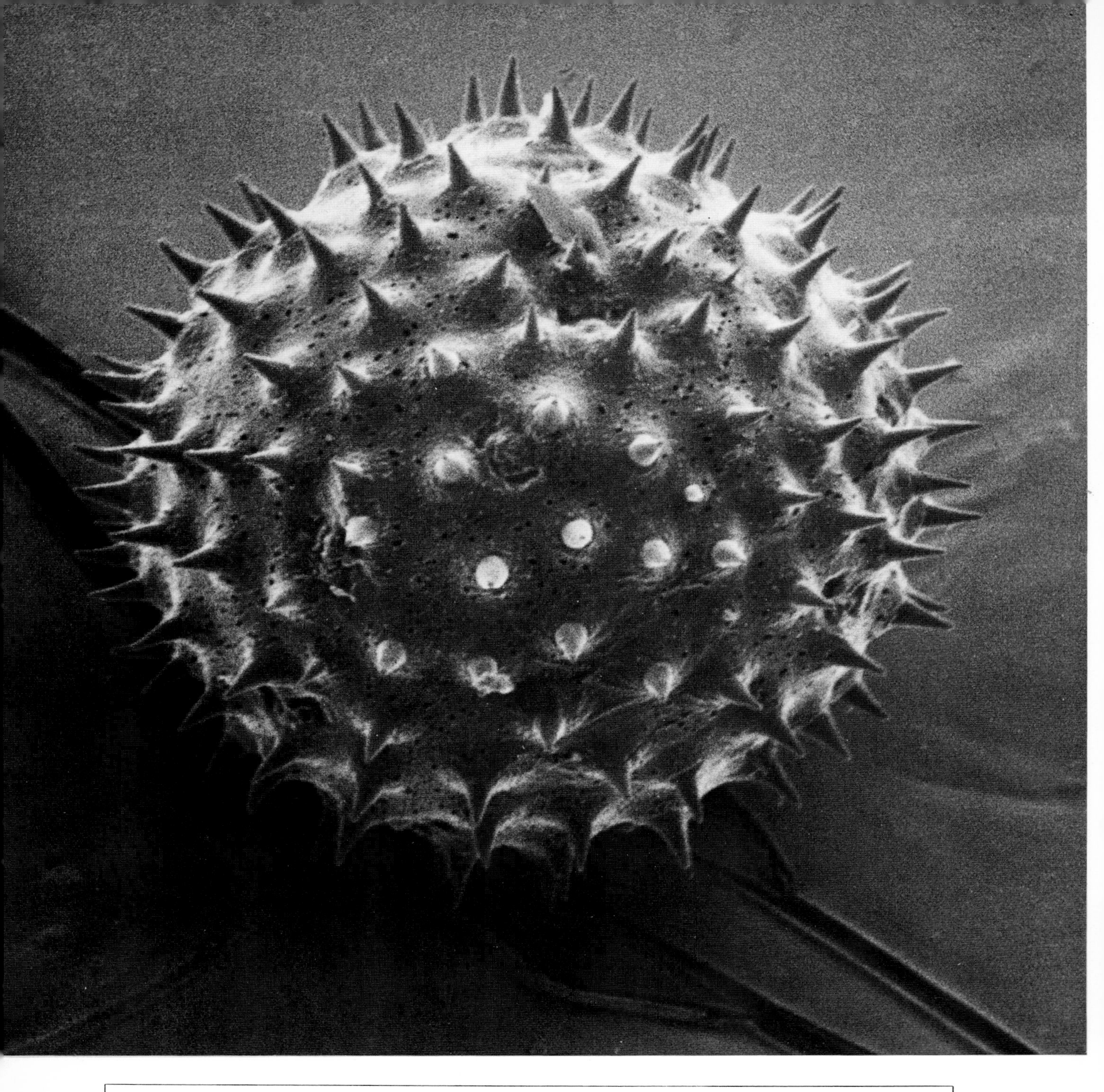

Plate 84 x2700 The Flower

Pollen grain of cotton. Dicotyledon.

Plate 85 x1250	The Flower

Pollen grain of maize. Monocotyledon.

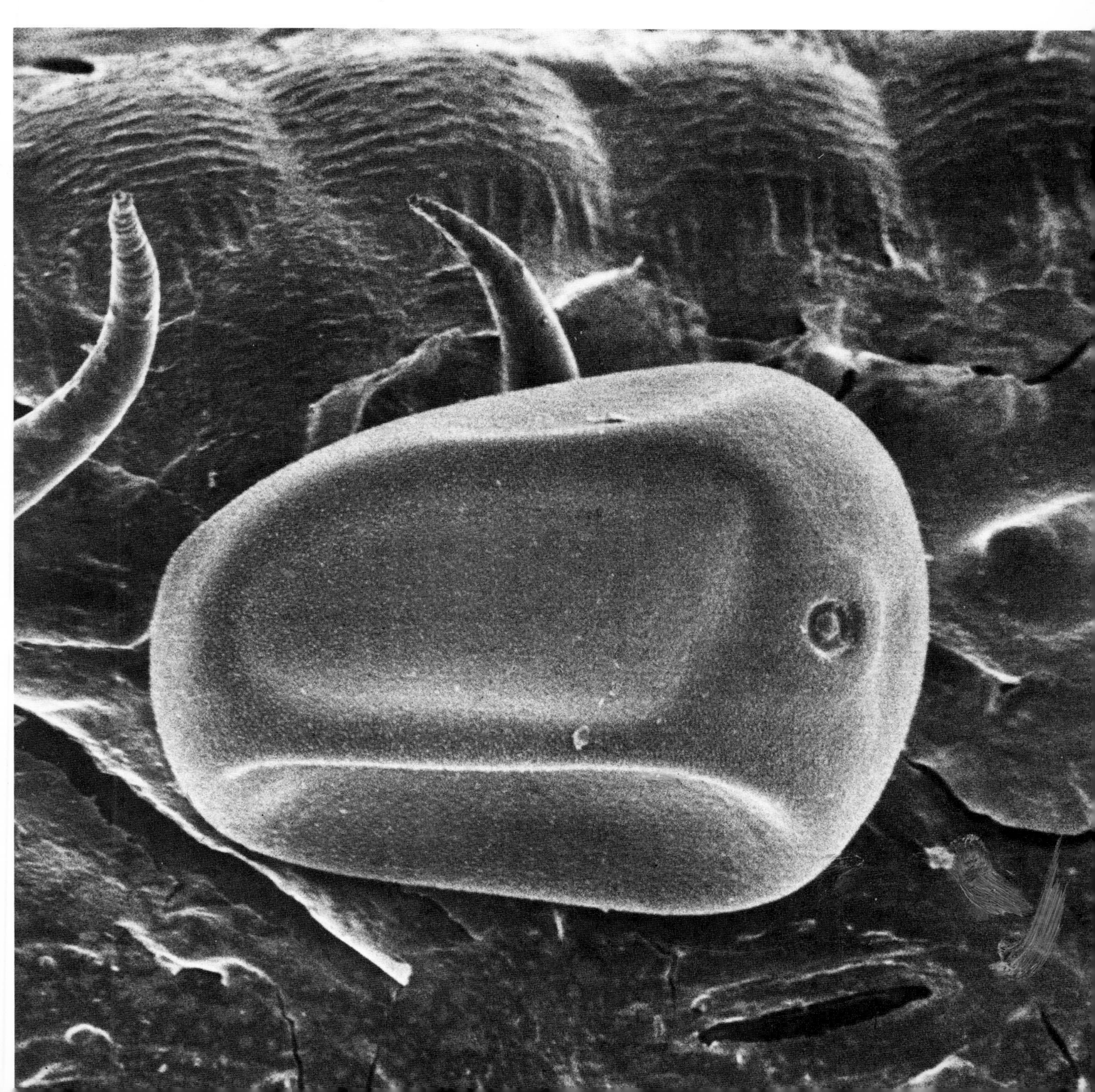

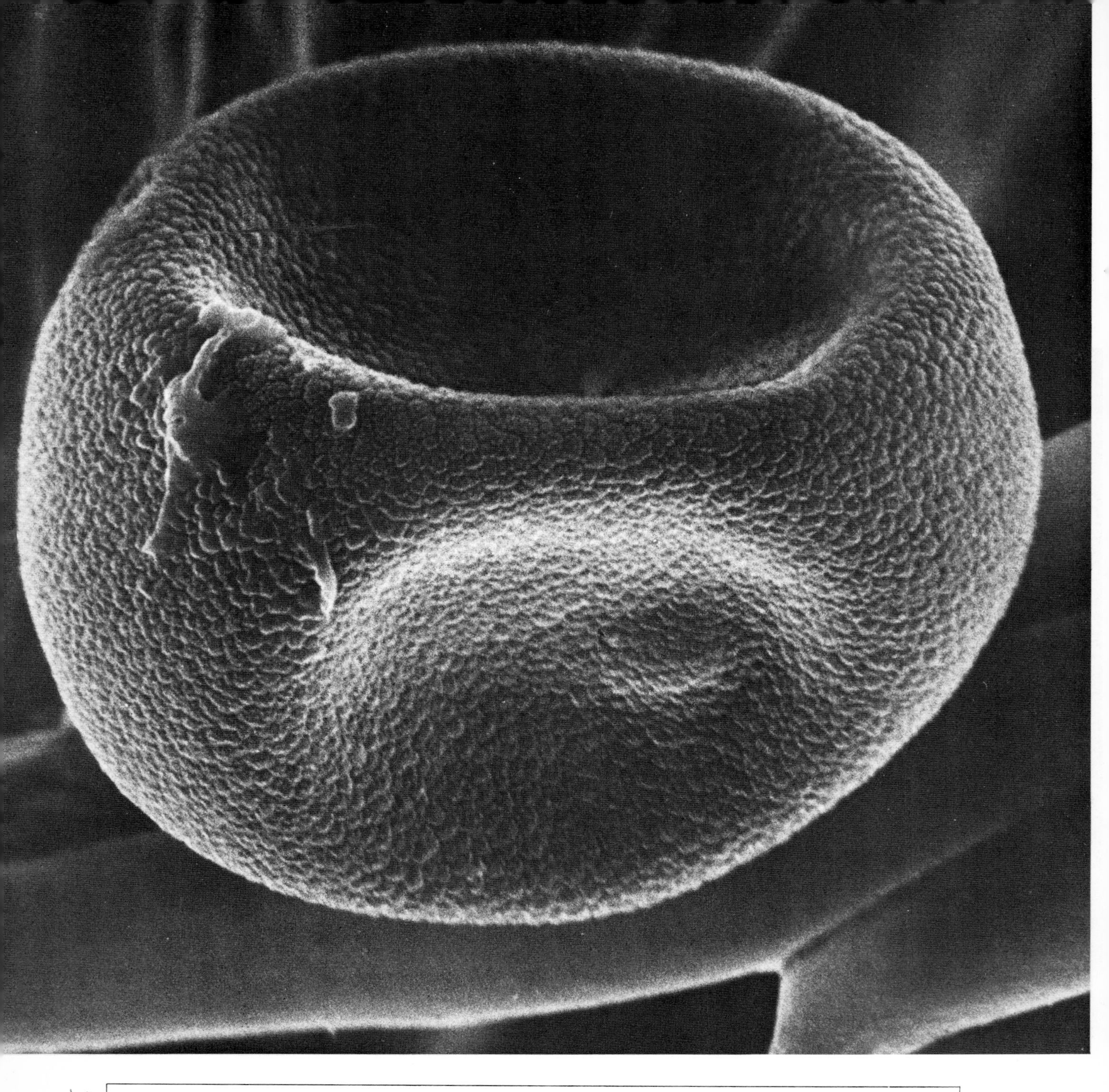

X **Plate 86** x1500 The Flower

Pollen grain of paspalum. Monocotyledon.

Plate 87 x4400 | The Flower

Pollen grain of pine. Gymnosperm.

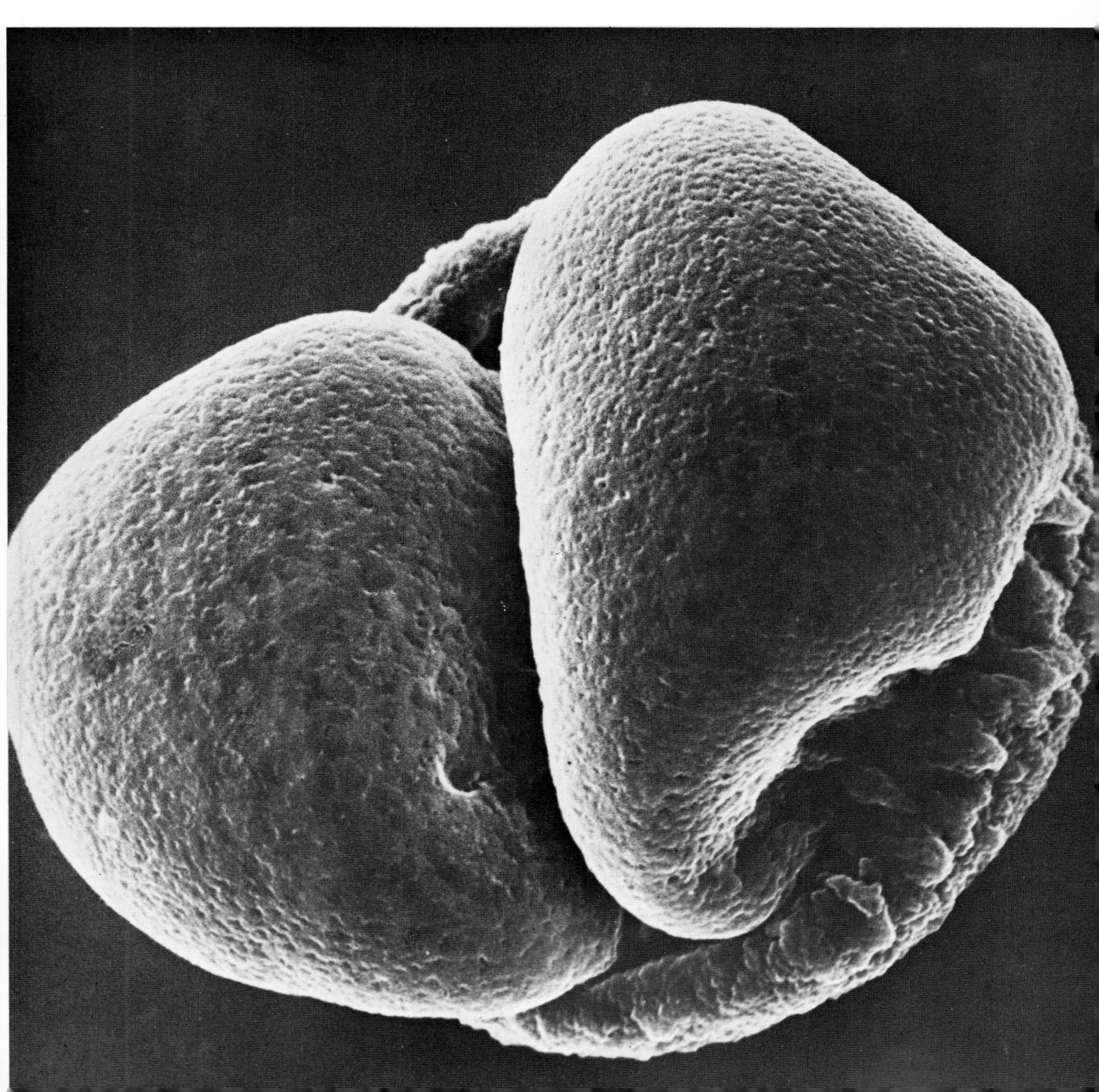

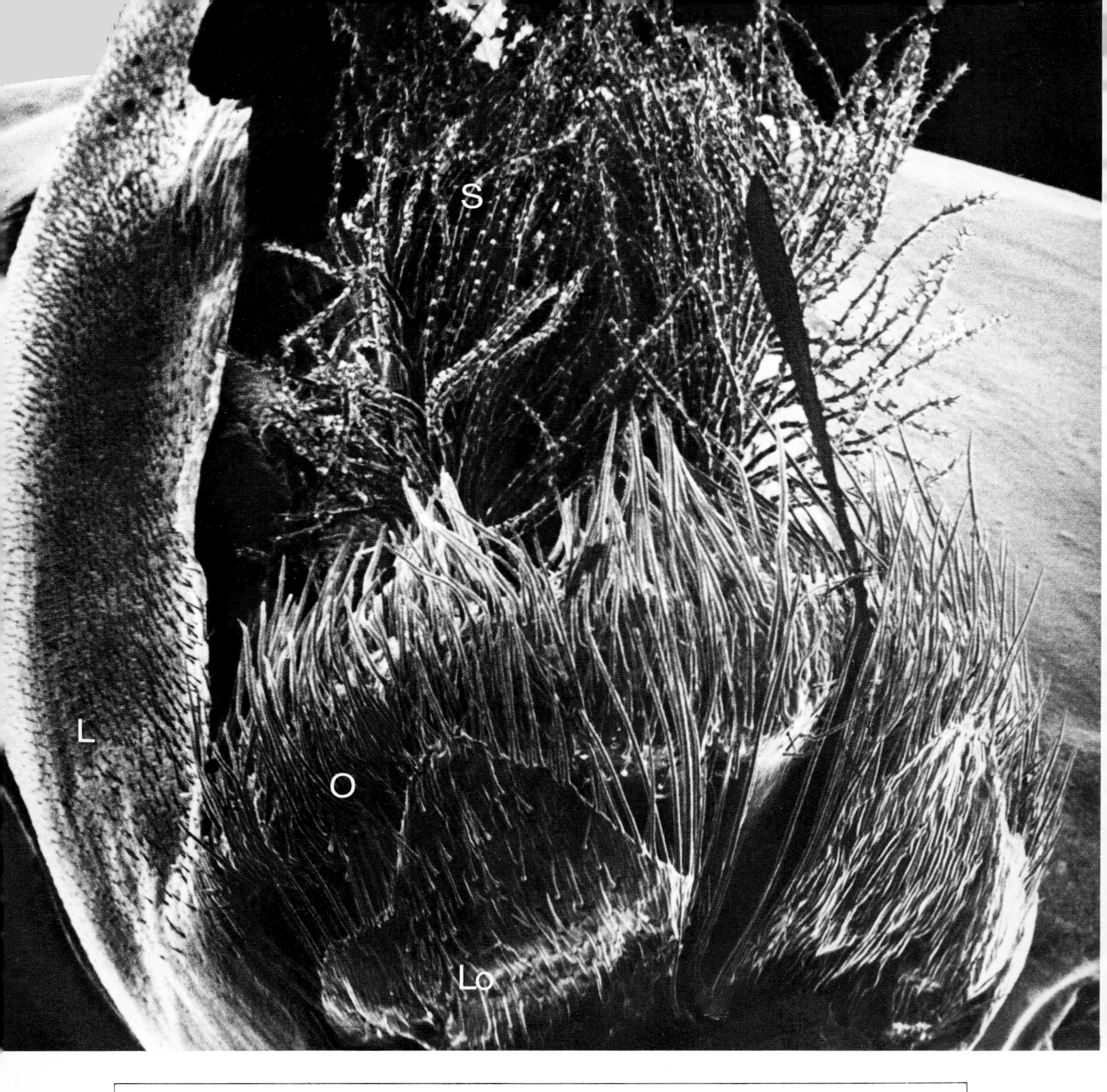

Plate 88 x50 The Flower

The flower of wheat with palea removed. To view the flower it was necessary to remove one of the bracts, the palea, which leaves the lemma (L) in its normal position. The bracts open naturally at anthesis by swelling action of the two scales of lodicules (Lo) at the base of the ovary. The ovary (O) is covered in hairs and the feathery stigmas (S) arise from the top of the ovary.

Fertilisation and Seed Development

Plate 89 x220	Fertilisation and Seed Development

Gynoecium from a flower of alyssum. The gynoecium is made up of a pollen trapping part at the top, the stigma, a fertile portion in the rounded base, the ovary, and these two structures are connected by the style. The stigma, style and ovary are covered by epidermal cells with a cuticle and stomata-like cells are present on the ovary. The ovules are attached to the inside wall of the ovary in a region termed the placenta. Upon fertilisation and subsequent development the ovules become the seed and the ovary the fruit.

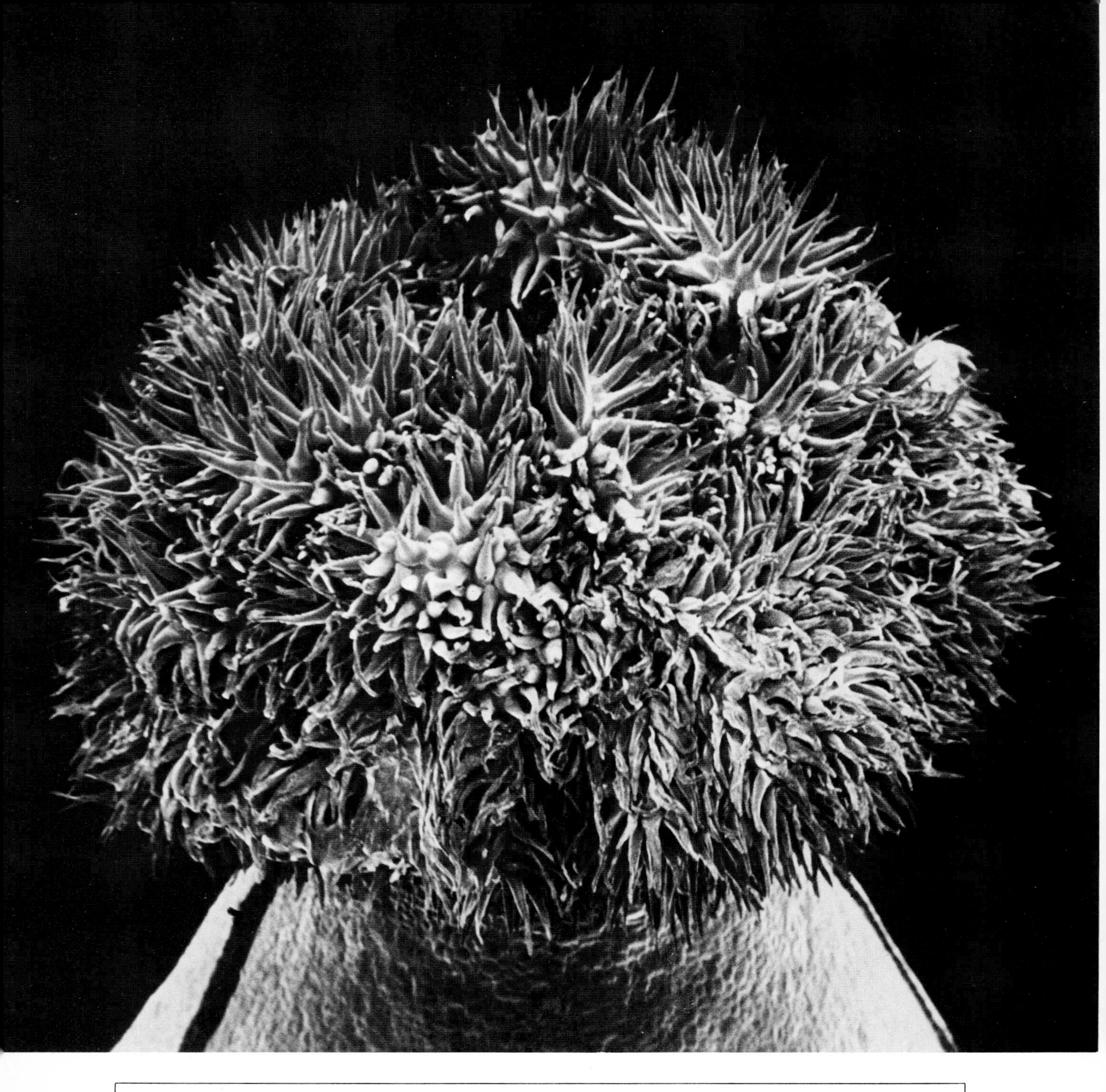

Plate 90 x120 — Fertilisation and Seed Development

Stigma from a flower of daphne *(Daphne sp.).* The stigma is the structure that is responsible for trapping pollen as the first stage in the fertilisation process.

Plate 91 x1900 Fertilisation and Seed Development

Stigma from a flower of primrose *(Primula sp.).* The epidermal cells making up the stigma are elongated into papillae and contain glandular tissue that secretes a sugary substance that helps trap the pollen grains, as well as assisting their germination.

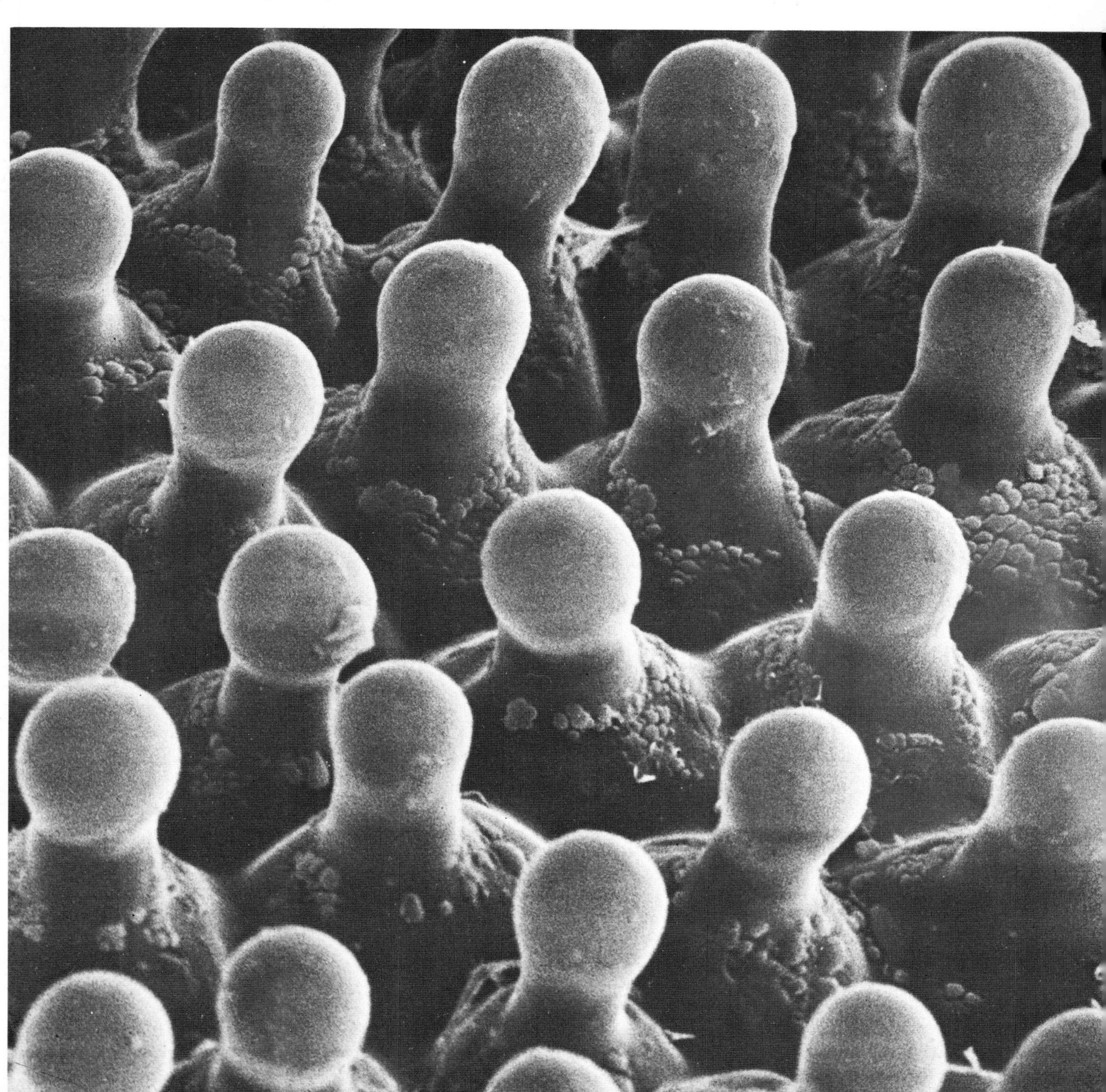

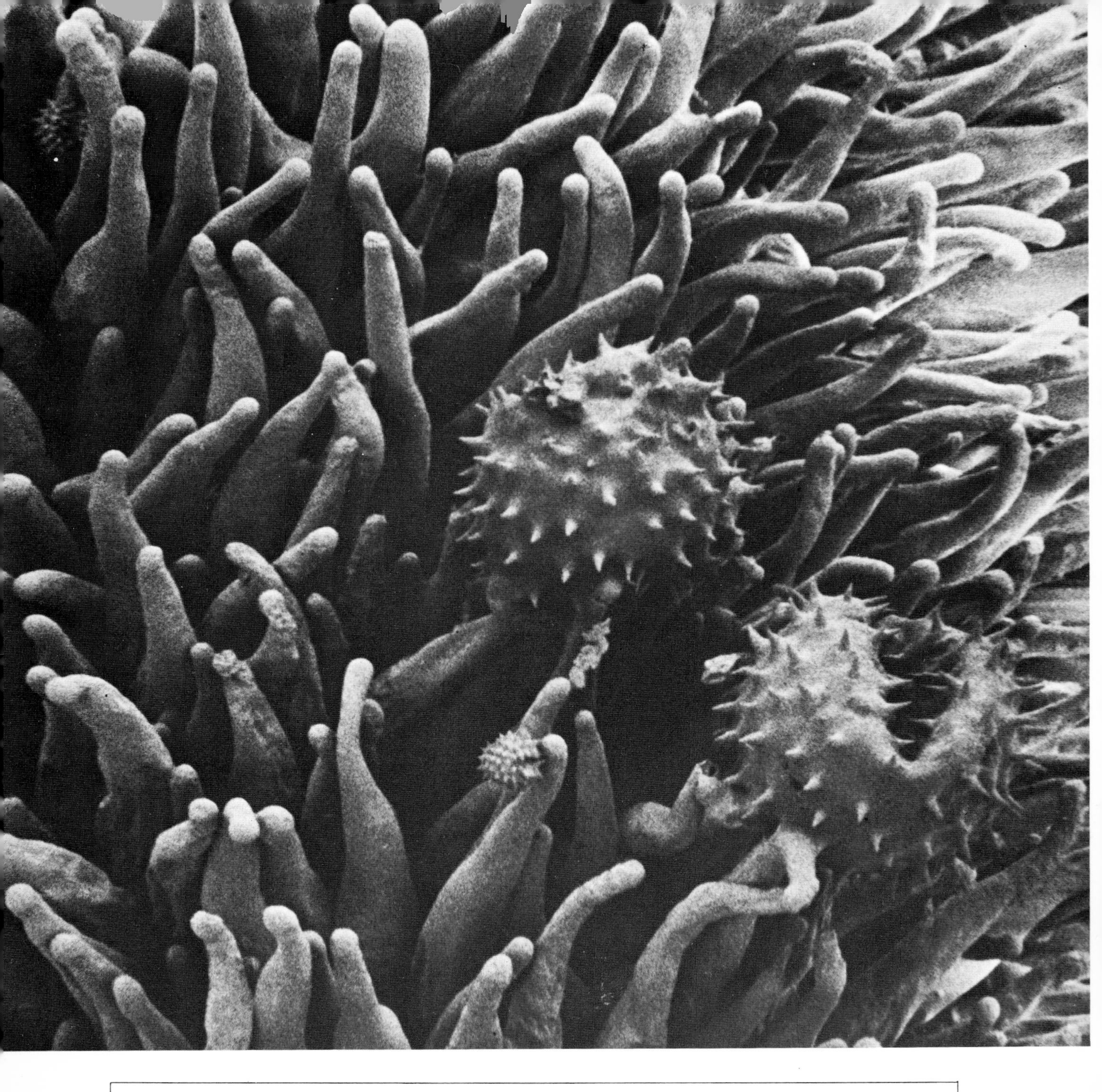

Plate 92 x570 Fertilisation and Seed Development

Pollen grains on the stigma of a cotton flower. There are two sizes of pollen grains shown in this photograph, the smaller ones which are sterile and the larger grains which have germinated. The indentation in the larger pollen grain indicates that the reserve substances in the grain have been used up in producing the pollen tube which in this example has penetrated the stigma. The pollen tube grows between the stigmatic papillae and into the tissue of the style. Generally the tube moves between the cells of the style and towards the ovary under the guidance of chemical attractants. The spines on the pollen grain are to assist the pollen grain being caught by the stigma.

Plate 93 x50 — Fertilisation and Seed Development

The stigma from a flower of gomphrena. Millions of pollen grains are produced by a flower but very few of these ever reach the stigma. Much pollen is lost from the flower completely and other grains fall on parts of the flower such as the sepals and petals and therefore do not take part in reproduction. It can be seen from the structure of the stigma of gomphrena shown in this picture that the chance of a pollen grain falling onto and being retained by stigma is small.

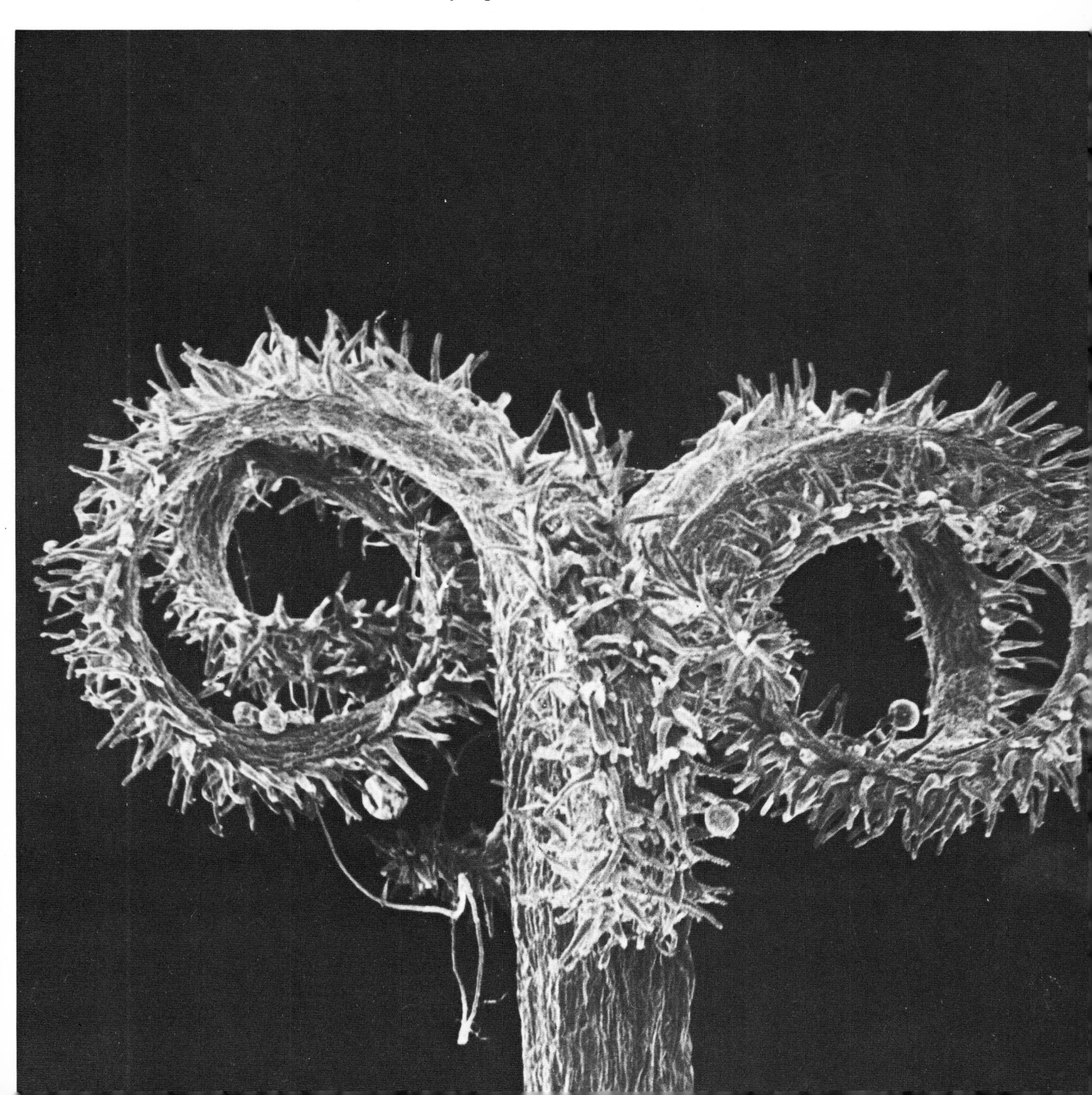

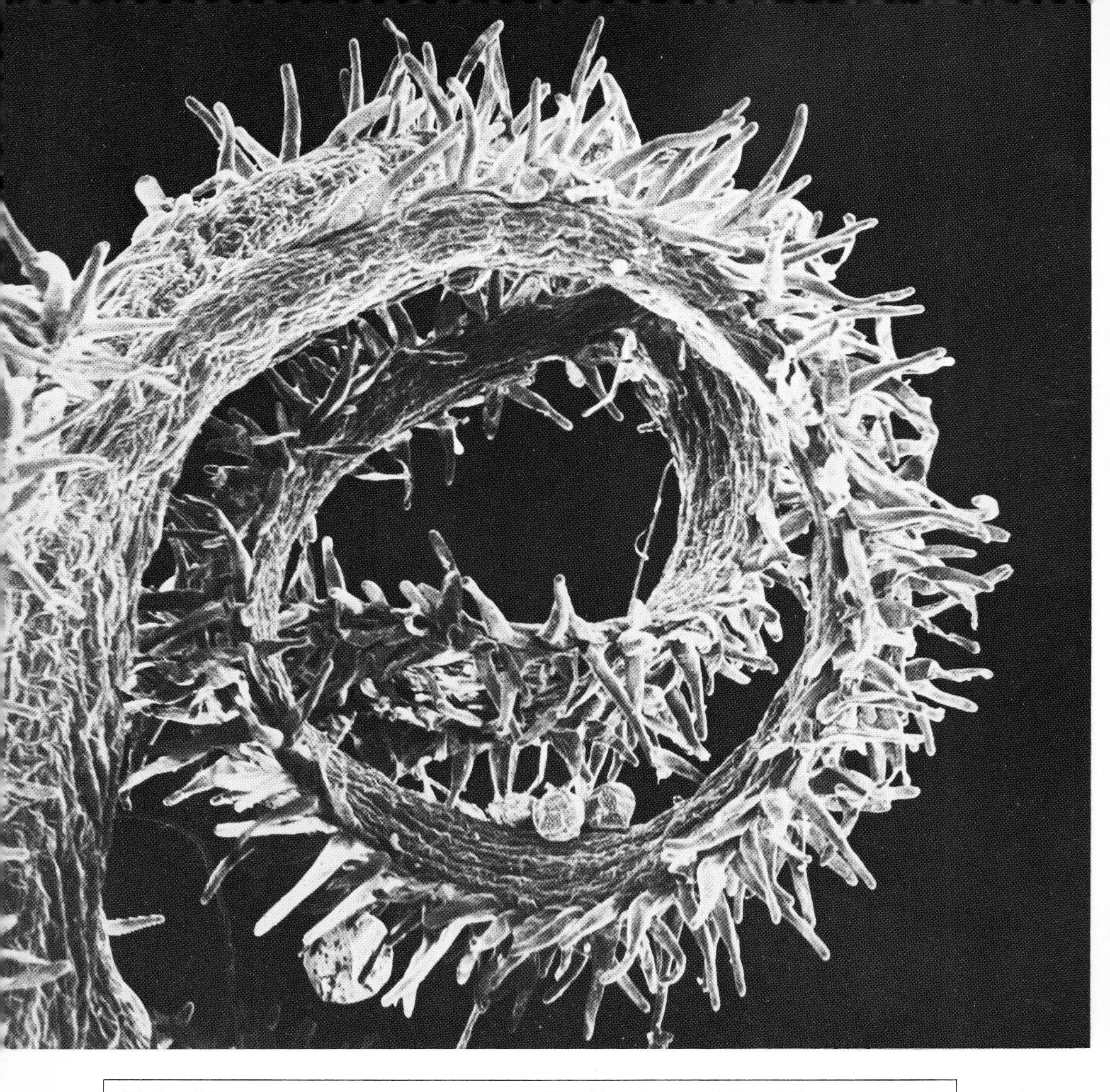

Plate 94 x100 Fertilisation and Seed Development

Pollen grains on the stigma of gomphrena. Once the pollen grains land on the moist stigmatic surface they take up water and germinate, sometimes within five minutes. The sugary substance secreted by the epidermal cells of the stigma assists in germination, and possibly provides a source of nutrients to the pollen tube. Most pollen grains only produce one pollen tube but in some species up to 14 pollen tubes have been known to be produced from a single grain. The pollen tube usually only lasts until fertilisation is completed.

Plate 95 x2300 Fertilisation and Seed Development

Pollen grains on the stigma of paspalum. Pollen grains in paspalum are windborne and only a few of the millions produced are actually involved in fertilisation. Hairs are often associated with the stigma of grasses to facilitate the trapping of the pollen and to make the process of capture of the grains more efficient. These hairs and stigma may be distant from the ovary and in maize it has been shown that some pollen tubes grow up to 50 cm. Over these long distances the tube obtains nutrients from the host plant and does not rely solely on the reserve material in the grain. The rate at which pollen tubes grow is variable and may be as high as 35 mm per hour.

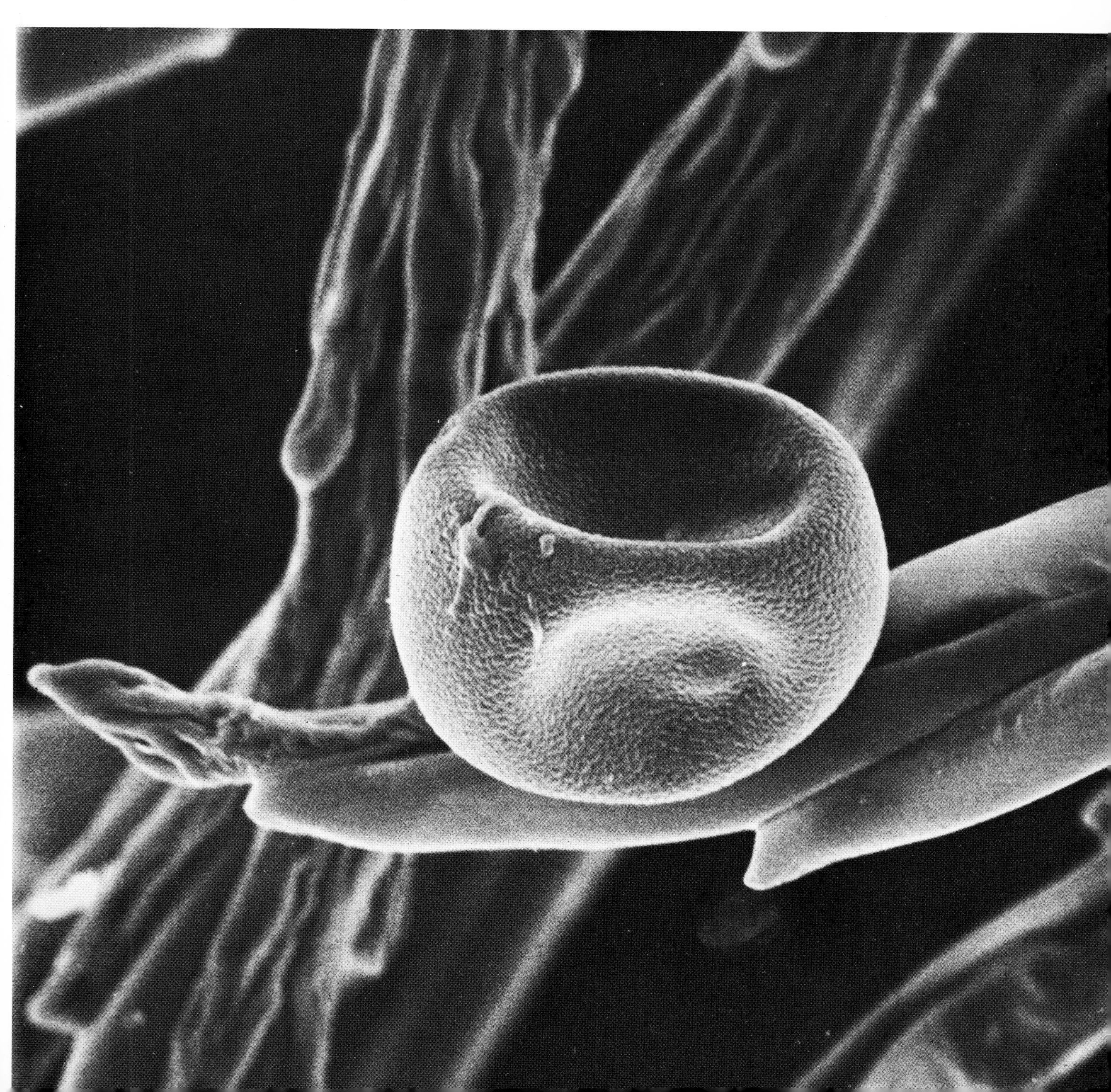

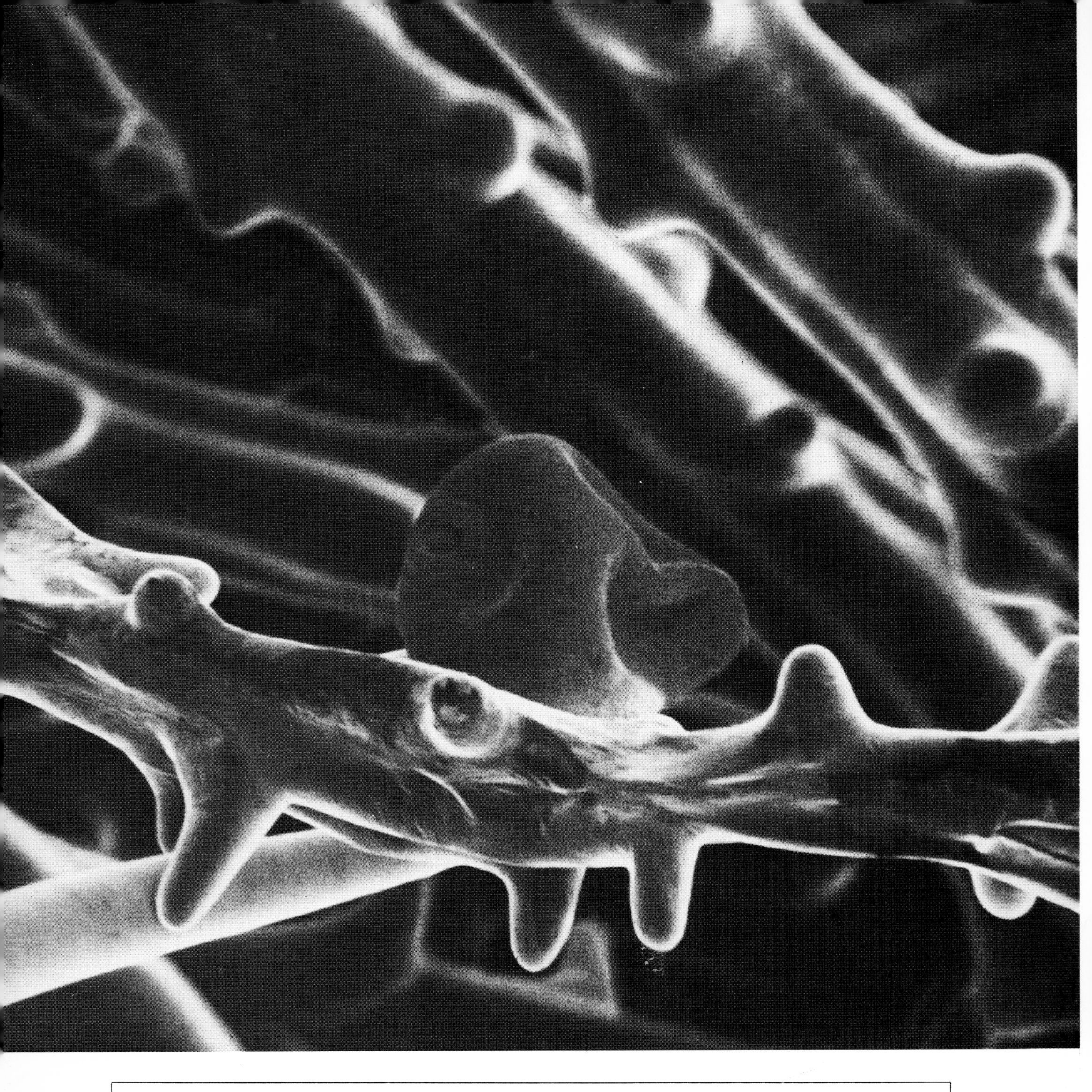

Plate 96 x1300 Fertilisation and Seed Development

Pollen grains on the stigma of wheat. In this photograph a germ pore is shown at one end of the pollen grain. Pollen consists of internal material or intine and an external coat of exine which has sporopollenin impregnated on the wall. The resultant tough external coat can resist decay by most organisms or by the environment and pollen grains have survived through geological time as fossils. Fertilisation occurs when the germ tube from the pollen grain arrives at the top of the ovary, enters the ovary through the micropyle and finally the pollen tube discharges its contents into the embryo sac. The time taken between germination of the pollen grain and fertilisation may be as short as 15 minutes or as long as 20 weeks.

Plate 97 x60 — Fertilisation and Seed Development

Fruit and receptacle of *Felicia ameloides.* After fertilisation the ovule develops into a seed which in this example is covered by a dry ovary wall and called the fruit. In some other plants the ovary can be stimulated, about the same time as fertilisation, into producing a fleshy fruit and in some cases a fruit may develop without fertilisation. By this stage of development the sepals, petals and stamens have withered and died. The growth and development of the seed of all plants is dependent on water, nutrients and carbon from the parent plant. As shown in this photograph the base of the seed is moulded into the receptacle and provides a close contact between the seed and the parent plant.

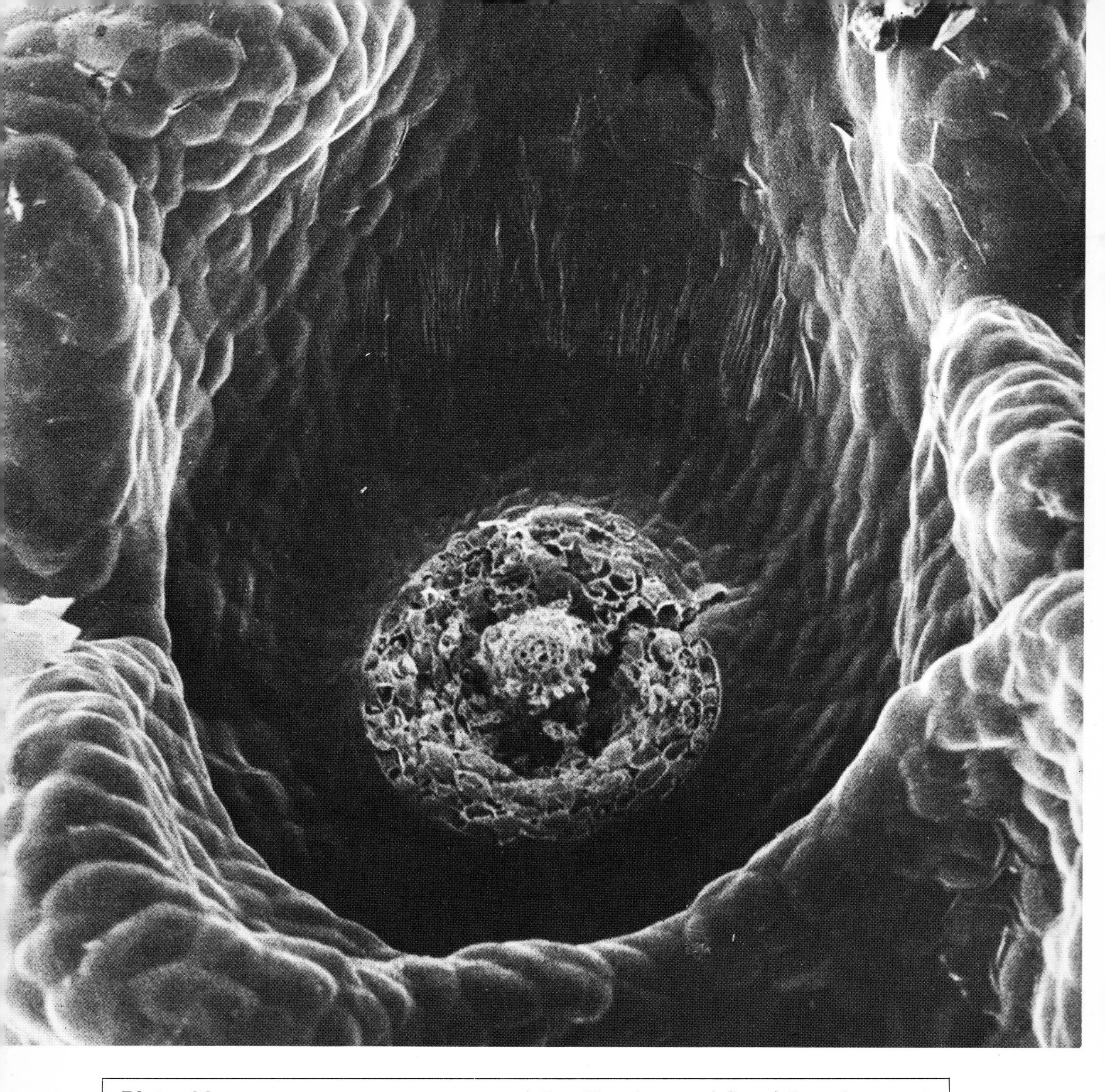

Plate 98 x640 Fertilisation and Seed Development

Single receptacle of *Felicia ameloides.* The vascular connections between the seed and receptacle can be seen in the base of this receptacle. The connection includes xylem and phloem with some extra tissue which acts as support. When the seed is mature the connective tissue degenerates and finally the seed is severed from the plant. Plants have developed numerous mechanisms to efficiently disperse the seeds away from the parent plant. As they dry out fruits twist and eject the seeds with an explosive force while other seeds have developed projections to assist dispersal by agents such as wind, animals and water.

Plate 99 x40 — Seeds

Seed case of burr clover *(Medicago hispida).* The case surrounding the seed of burr clover protects the seed during development, but has been modified to assist seed dispersal. As the name of the clover suggests, there are burrs or hooks over the surface of the seed case which readily become entangled in the coats of many animals. The seeds can thus be carried over long distances.

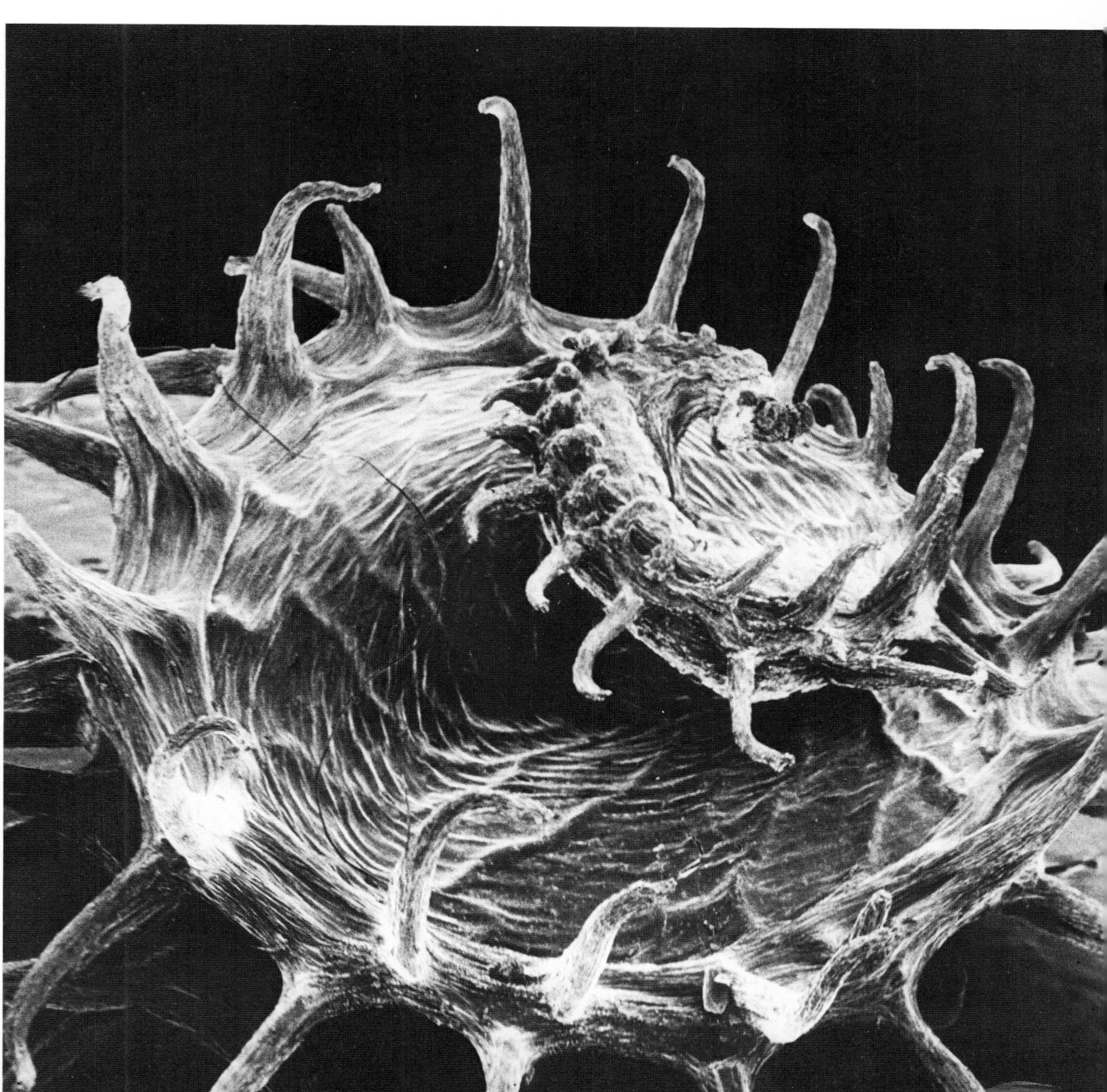

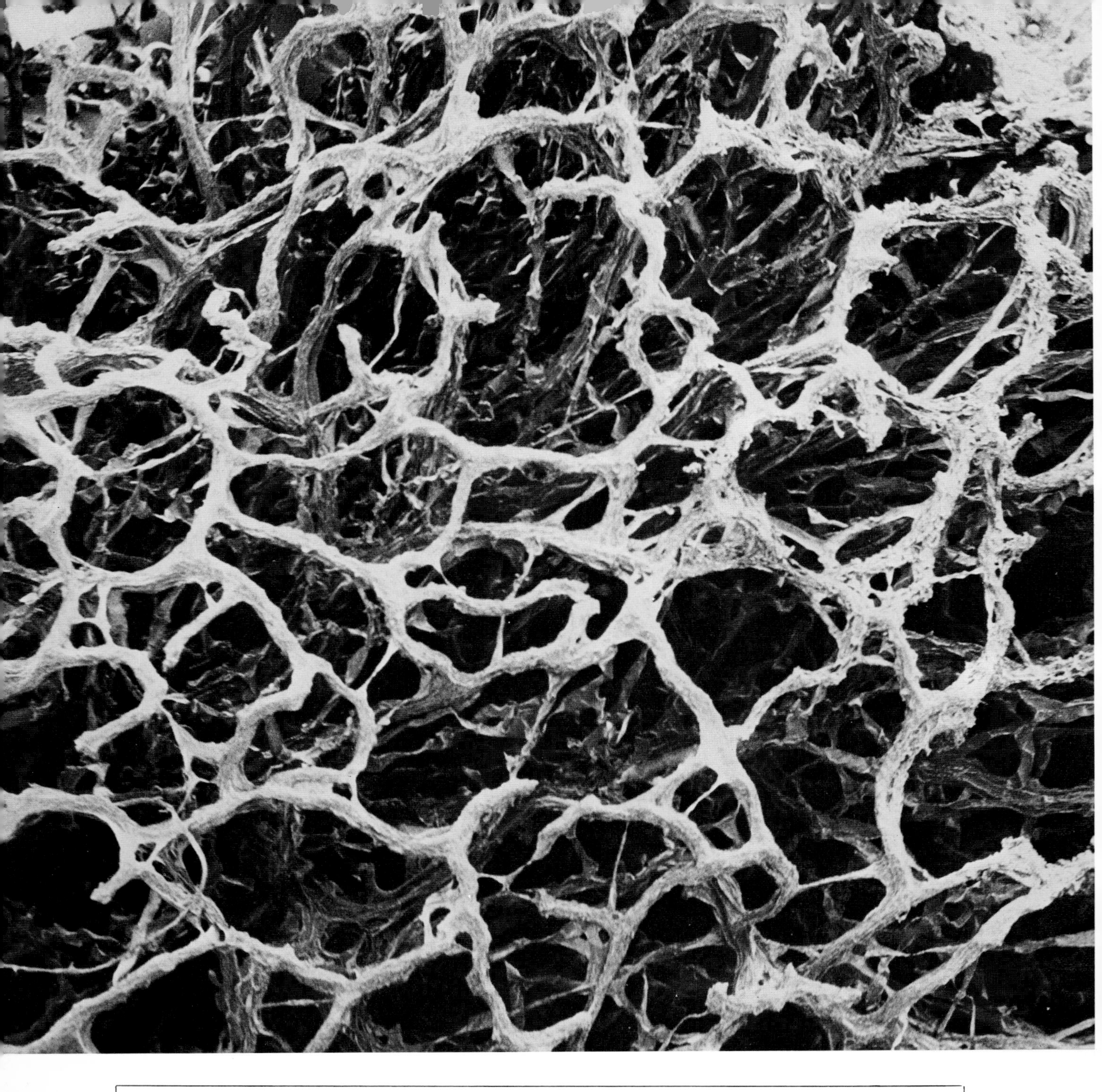

Plate 100 x50 — Seeds

Seed case of saltbush. The seed in this species is surrounded by a network of fibres, the remnants of vascular bundles similar to those in the bundle sheath of a saltbush leaf. While the seed is developing, photosynthesis in this layer of tissue round the seed will contribute carbon to the seed. After the seed is lost from the plant this sponge-like network separates the seed from the soil by about 2 mm. Saltbush occurs in arid regions but for the seed to germinate a sufficiently prolonged or heavy shower of rain is necessary—to overcome the effect of the seed case in keeping the seed off the ground and therefore dry. Without this seed case, spasmodic light showers of rain could cause the seed to germinate but would provide insufficient water to allow growth to proceed normally.

Plate 101 x50 Seeds

Seed of *Galinsoga parviflora.*

Plate 102 x310	Seeds

Pappas on a *Galinsoga parviflora* seed.

Plate 103 x280 Seeds

Seed of an *Epilobium sp.*

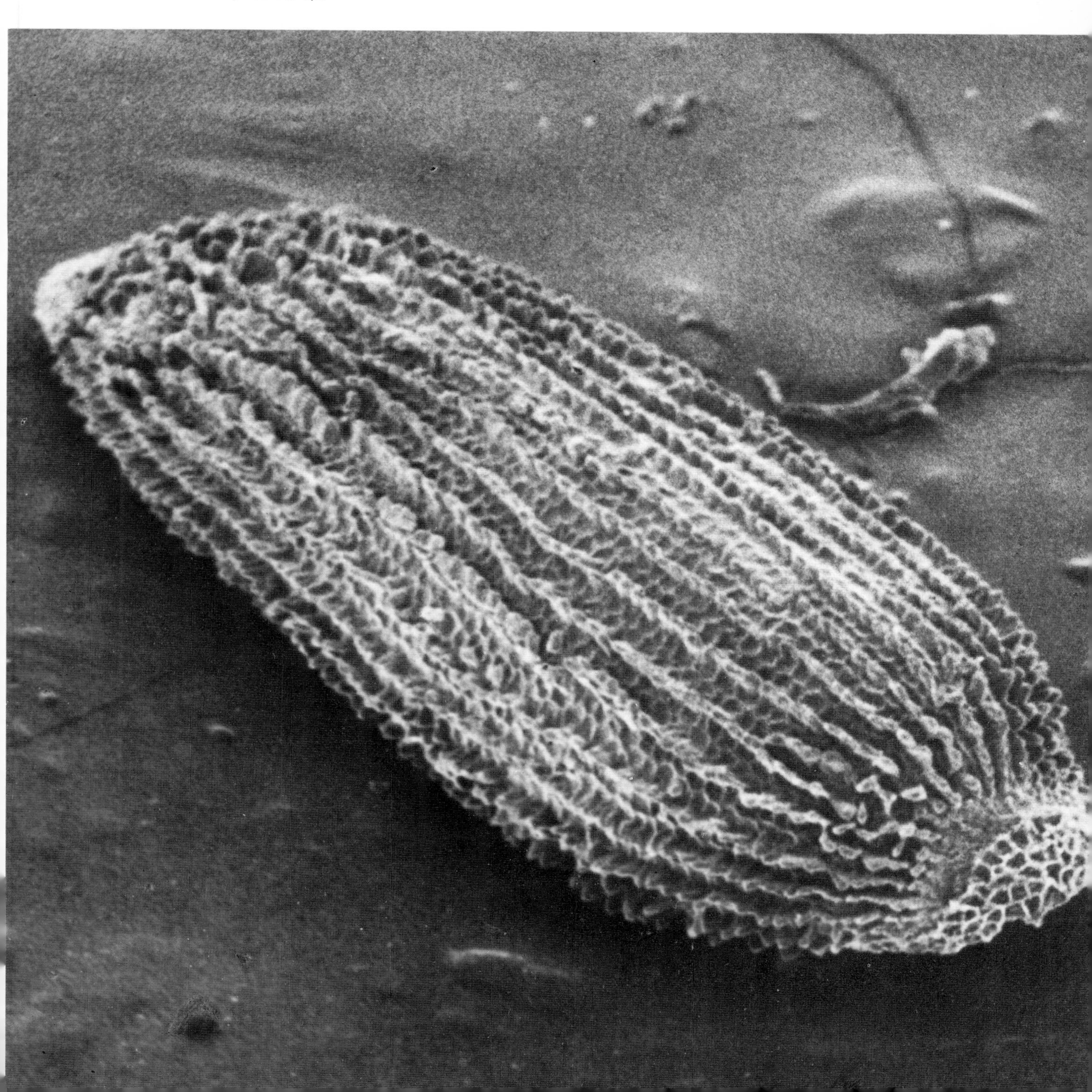

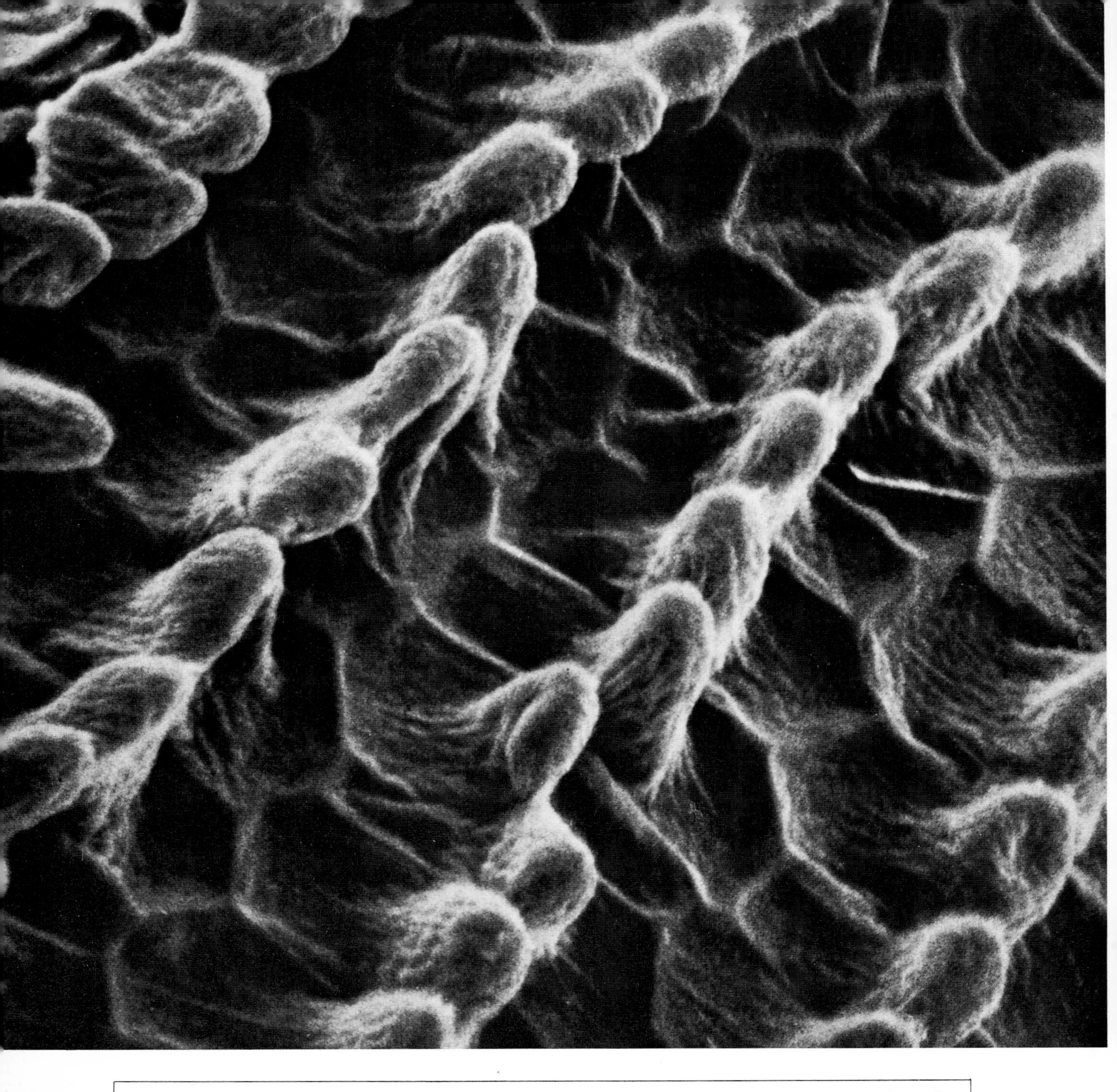

Plate 104 x2040	Seeds

Seed of an *Epilobium sp.*

Plate 105 x130	Seeds

Seed of bladder campion *Silene cucubalus.*

Bibliography

Books which may be consulted for details on many aspects of plant anatomy or physiology:

Plant Anatomy 2nd Edition. K. Esau. Published 1965 by Wiley, New York.

Plant Structure and Development. T. P. O'Brien and Margaret E. McCully. Published 1969 by Macmillan, London.

Botany: An Introduction to Plant Biology 4th Edition. T. E. Weir, C. R. Stocking and M. G. Barbour. Published 1970 by Wiley, New York.

Biology of Plants. P. H. Raven and Helena Curtis. Published 1970 by Worth Publishers, New York.

Anatomy of Seed Plants. K. Esau. Published 1960 by Wiley, New York.

Photosynthesis and Photorespiration. M. D. Hatch, C. B. Osmond and R. O. Slatyer. Published 1971 by John Wiley Interscience, New York.